SCIENCE & Life Issues

SCIENCE & Life Issues

SCIENCE
EDUCATION FOR
PUBLIC
UNDERSTANDING
PROGRAM

S E P U P

UNIVERSITY OF CALIFORNIA AT BERKELEY

LAWRENCE HALL OF SCIENCE **LHS**

LaB-aiDS
INCORPORATED

RONKONKOMA, NEW YORK

This book is part of SEPUP's middle school science course sequence:

Issues and Earth Science

Studying Soil Scientifically
Rocks and Minerals
Erosion and Deposition
Plate Tectonics
Weather and Atmosphere
The Earth in Space
Exploring the Solar System

Science and Life Issues

Studying People Scientifically
Body Works
Micro-Life
Our Genes, Our Selves
Ecology
Evolution
Using Tools and Ideas

Issues and Physical Science

Studying Materials Scientifically
The Chemistry of Materials
Water
Energy
Force and Motion

Additional SEPUP instructional materials include:
CHEM-2 (Chemicals, Health, Environment and Me): Grades 4–6
SEPUP Modules: Grades 7–12
Science and Sustainability: Course for Grades 9–12

 This material is based upon work supported by the National Science Foundation under Grant No. 9554163. Any opinions, findings, and conclusions or recommendations expressed in this material are those of the authors and do not necessarily reflect the views of the National Science Foundation.

SEPUP
Lawrence Hall of Science
University of California at Berkeley
Berkeley CA 94720-5200

e-mail: sepup@berkeley.edu
Website: www.sepuplhs.org

Published by:

17 Colt Court
Ronkonkoma NY 11779
Website: www.lab-aids.com

A Letter to SALI Students

As you examine the activities in this book, you may wonder, "Why does this book look so different from other science books I've seen?" The reason is simple: it is a different kind of science program, and only some of what you will learn can be seen by leafing through this book!

Science and Life Issues, or *SALI,* uses several kinds of activities to teach science. For example, you will design and conduct an experiment to investigate human responses. You will explore a model of how species compete for food. And you will play the roles of scientists learning about the causes of infectious disease. A combination of experiments, readings, models, debates, role plays, and projects will help you uncover the nature of science and the relevance of science to your interests.

You will find that important scientific ideas come up again and again in different activities. You will be expected to do more than just memorize these concepts: you will be asked to explain and apply them. In particular, you will improve your decision-making skills, using evidence and weighing outcomes to decide what you think should be done about scientific issues facing society.

How do we know that this is a good way for you to learn? In general, research on science education supports it. In particular, the activities in this book were tested by hundreds of students and their teachers, and they were modified on the basis of their feedback. In a sense, this entire book is the result of an investigation: we had people test our ideas, we interpreted the results, and we revised our ideas! We believe the result will show you that learning more about science is important, enjoyable, and relevant to your life.

SALI Staff

SCIENCE & LIFE ISSUES PROJECT

Director (2003–2008): Barbara Nagle
Director (1995–2002): Herbert D. Thier

CONTRIBUTORS/DEVELOPERS

Barbara Nagle
Manisha Hariani
Herbert D. Thier
Asher Davison
Susan K. Boudreau
Daniel Seaver
Laura Baumgartner

TEACHER CONTRIBUTORS

Kathaleen Burke
Richard Duquin
Donna Markey

CONTENT AND SCIENTIFIC REVIEW

Jim Blankenship, Professor and Chairman of Pharmacology, School of Pharmacy, University of the Pacific, Stockton, California *(Studying People Scientifically* and *Micro-Life)*

Gary R. Cutter, Director of Biostatistics, AMC Cancer Research, Denver, Colorado *(Studying People Scientifically* and *Body Works)*

Peter J. Kelly, Emeritus Professor of Education and Senior Visiting Fellow, School of Education, University of Southampton, Southampton, England *(Complete course)*

Eric Meikle, National Center for Science Education, Oakland, California *(Evolution)*

Deborah Penry, Assistant Professor, Department of Integrative Biology, University of California at Berkeley, Berkeley, California *(Complete course)*

Arthur L. Reingold, Professor, Department of Public Health Biology and Epidemiology, University of California at Berkeley, Berkeley, California *(Micro-Life)*

RESEARCH ASSISTANCE

Marcelle Siegel, Leif Asper

PRODUCTION

Project coordination: Miriam Shein
Production and composition: Seventeenth Street Studios
Cover concept: Maryann Ohki
Photo research and permissions: Sylvia Parisotto
Editing: WordWise

Field Test Centers

The classroom is SEPUP's laboratory for development. We are extremely appreciative of the following center directors and teachers who taught the program during the 1998–99 and 1999–2000 school years. These teachers and their students contributed significantly to improving the course.

REGIONAL CENTER, SOUTHERN CALIFORNIA
Donna Markey, *Center Director*
Kim Blumeyer, Helen Copeland, Pat McLoughlin, Donna Markey,
Philip Poniktera, Samantha Swann, Miles Vandegrift

REGIONAL CENTER, IOWA
Dr. Robert Yager and Jeanne Bancroft, *Center Directors*
Rebecca Andresen, Lore Baur, Dan Dvorak, Dan Hill, Mark Kluber, Amy Lauer,
Lisa Martin, Stephanie Phillips

REGIONAL CENTER, WESTERN NEW YORK
Dr. Robert Horvat, *Center Director*
Kathaleen Burke, Mary Casion, Dick Duquin, Eleanor Falsone, Lillian Gondree,
Jason Mayle, James Morgan, Valerie Tundo

JEFFERSON COUNTY, KENTUCKY
Pamela Boykin, *Center Director*
Charlotte Brown, Tara Endris, Sharon Kremer, Karen Niemann,
Susan Stinebruner, Joan Thieman

LIVERMORE, CALIFORNIA
Scott Vernoy, *Center Director*
Rick Boster, Ann Ewing, Kathy Gabel, Sharon Schmidt, Denia Segrest,
Bruce Wolfe

QUEENS, NEW YORK
Pam Wasserman, *Center Director*
Gina Clemente, Cheryl Dodes, Karen Horowitz, Tricia Hutter, Jean Rogers,
Mark Schmucker, Christine Wilk

TUCSON, ARIZONA
Jonathan Becker, *Center Director*
Peggy Herron, Debbie Hobbs, Carol Newhouse, Nancy Webster

INDEPENDENT
Berkeley, California: Robyn McArdle
Fresno, California: Al Brofman
Orinda, California: Sue Boudreau, Janine Orr, Karen Snelson
Tucson, Arizona: Patricia Cadigan, Kevin Finegan

Contents

UNIT B Body Works

UNIT C   Micro-Life

UNIT D Our Genes, Our Selves

UNIT F Evolution

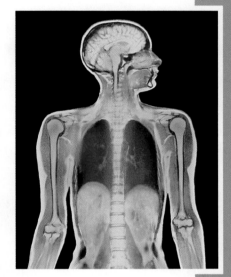

Studying People Scientifically

A

Unit A

MEDICAL RESEARCH

STUDY PARTICIPANTS NEEDED

Be part of a medical research study and earn up to $
a day. Your participation could help improve the qu
of life for people with various health condit
YOU CAN MAKE A DIFFERENCE! Study partic
who meet the following criteria are eligible:

- 18 to 68 years old
- male or female
- available weekends/weekd
- non-smokers

Studying People Scientifically

"**B**ut it's true. I heard about it on the news last night. Broccoli decreases your risk of having cancer," stated Alissa.

Eric rolled his eyes. "They used to say margarine was much better for you than butter, then it was butter is better—who knows what medical researchers are saying today?"

"I don't know, Eric, this sounded pretty convincing," replied Alissa. "They've done scientific studies and everything."

"I'm not going to eat it just because some doctor says I should," said Eric stubbornly.

• • •

What would you do if you were Eric? What factors might influence your view? How do you respond to everyday dilemmas, and on what basis do you make your decisions?

Scientists collect information and do studies to answer such questions. In this unit, you will learn some of the scientific principles and approaches used to study people.

We all solve problems every day—from little problems like what to wear to school, to bigger problems like whether to get a job or go to college. Different kinds of problems require different problem-solving approaches. One problem can often be solved in more than one way. What do you do when you are faced with a problem?

CHALLENGE

How can you and your partner solve a problem?

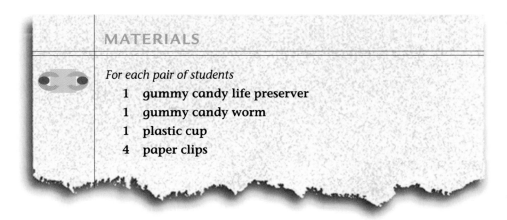

MATERIALS

For each pair of students

1 **gummy candy life preserver**
1 **gummy candy worm**
1 **plastic cup**
4 **paper clips**

PROCEDURE

1. Work with your partner to solve the problem below. A picture of the set-up is on the next page.

SAVE FRED!

Poor Fred! He was sailing along on a boat (your plastic cup) when a strong wind blew it upside-down. Fred (your candy worm) ended up on top of the upside-down boat. Unfortunately for Fred, his life preserver (your candy life preserver) is still trapped under the boat.

Your job is to place the life preserver firmly around Fred's body, but you must obey three rules:

1. Fred may not fall into the "sea" (onto the table) more than one time; if he does, Fred "drowns."

2. You may not injure him in any way.

3. You may use only the four paper clips to move Fred, the boat, and the life preserver. You may not touch anything except the paper clips.

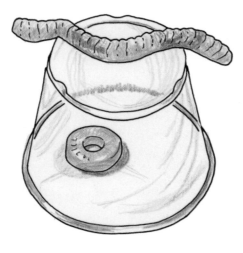

2. Work with your partner to record in your science notebook exactly what you did to save Fred. You may wish to draw a picture or a diagram to explain your procedure.

3. Explain your procedure to another two partners.

ANALYSIS

Each activity in this book asks Analysis Questions. These questions will guide you in your learning. You will not have to write your response to all of the questions nor will you have to answer all of them by yourself. Use the key below as well as directions from your teacher to find out exactly what to do.

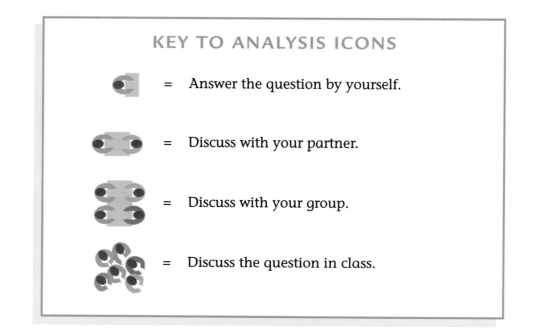

KEY TO ANALYSIS ICONS

= Answer the question by yourself.

= Discuss with your partner.

= Discuss with your group.

= Discuss the question in class.

1. You can solve problems in many different ways. In fact, you may use more than one way to solve a single problem. You can

   - develop a plan.
   - find a pattern.
   - draw a picture or a diagram.
   - act out the problem.
   - make a list.
   - guess and test.

   - work backward.
   - write an equation.
   - construct a table or graph.
   - simplify the problem.
   - use objects to model the problem.

   Which of these ways did you and your partner use to save Fred?

2. As a class, discuss the ways in which various groups of partners saved Fred. How were your problem-solving methods similar? How were they different?

3. **Reflection:** People face problems in their lives every day. What did you learn from this activity that you can use to solve other problems?

VIEW AND REFLECT

Scientists have common approaches to solving problems. Usually, people think of experiments as one way scientists investigate problems. What kinds of experiments are possible when you study human beings? How can you collect evidence in these situations? Begin to consider these issues as you watch the story of the disease called pellagra (puh-LAY-gra), which affected poor rural families of the South.

Whenever you see this icon, it means that you will find more information or a technology extension on this subject on the SALI page of the SEPUP website.

Poor families of the South were more likely to develop pellagra.

CHALLENGE

What are the common elements shared by all scientific problem-solving methods? How is science used to study people?

PROCEDURE

1. In order to prepare to watch the story on the video, first read Analysis Questions 1–4.

2. Your teacher will provide you with a student sheet or ask you to prepare a table like Table 1 on the next page to record your notes during the video.

3. Watch the story of pellagra on the video, *A Science Odyssey*: "Matters of Life and Death."

Table 1: Notes on The Pellagra Story

What was the problem of pellagra?

_____

What did people think caused pellagra?

1.

2.

3.

4.

_____

What evidence did Dr. Goldberger observe or collect about pellagra?

_____

What did Dr. Goldberger conclude about the cause of pellagra?

_____

ANALYSIS

1. a. What was the first step in Dr. Goldberger's research into pellagra? Explain why this step was important in developing his hypothesis.

b. During this first step in his research, what evidence did Dr. Goldberger find that suggested that pellagra was not caused by germs?

2. a. What was Dr. Goldberger's hypothesis about the cause of pellagra?

b. What did he do to provide evidence of the relationship between pellagra and nutrition? Be sure to explain *how* his research provided evidence that supported or disproved his hypothesis.

c. How could he have provided more convincing evidence of the relationship between pellagra and nutrition?

Dr. Joseph Goldberger

3. Why didn't people believe Dr. Goldberger's conclusion about the cause of pellagra? Give two reasons.

4. Compare the steps of the traditional scientific method to the steps Dr. Goldberger followed to investigate pellagra. How were the steps the same? How were the steps different?

5. To investigate his hypothesis, Dr. Goldberger had prisoners volunteer to be fed a poor diet; as a result, 7 out of 11 prisoners developed pellagra. What do you think about Dr. Goldberger's decision to experiment on people? Support your answer with evidence and identify the trade-offs of your decision.

   Hint: To write a complete answer, first state your opinion. Provide two or more pieces of evidence that support your opinion. Then consider all sides of the issue and identify the trade-offs of your decision.

6. **Reflection:** How do people in different careers solve problems? Scientists, plumbers, engineers, auto mechanics, nurses, teachers, and many other workers solve problems. Choose two careers that interest you. Describe the kind of problems you think people face in these careers. Describe how you think they solve them.

Collecting evidence is an important part of science. One way to collect evidence is to conduct experiments as Dr. Goldberger did. Products such as medicines are tested by volunteers before they are made available to the public. These tests are known as **clinical trials**.

A CLINICAL TRIAL

Imagine that you suffer from severe headaches several times a month. These headaches are so painful that you can't read, listen to music, or watch television. Regular headache medicines don't work very well for you. One day, you complain to your doctor about your headaches. She tells you that the local medical school is conducting clinical trials of a new headache medicine. She asks if you would like to volunteer to be a part of this trial. Hoping for relief, you say yes.

Since medicines cannot be tested in the classroom, you will participate in a simulation of a clinical trial. In this simulation, differences in taste will equal differences in response to the medicine. Figure 1, "Clinical Trial of A Headache Medicine," on the next page, explains this simulation.

CHALLENGE

How are medicines tested during a clinical trial?

Figure 1: Clinical Trial of A Headache Medicine

The taste of the yellow lemon drink represents a headache.

The taste of the pink lemon drink represents the medicine taken to treat your headache.

If the pink lemon drink tastes the same as the yellow, there is no change in your headache.

If the pink lemon drink tastes better than the yellow, your headache is gone!

If the pink lemon drink tastes worse than the yellow, your headache is gone, but you experience side effects.

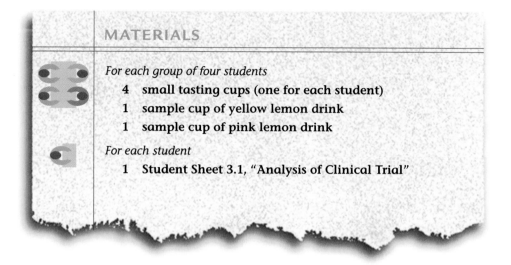

MATERIALS

For each group of four students

  4 small tasting cups (one for each student)

  1 sample cup of yellow lemon drink

  1 sample cup of pink lemon drink

For each student

  1 Student Sheet 3.1, "Analysis of Clinical Trial"

SAFETY

Never taste materials or eat or drink in science class unless specifically told to do so by your teacher. Be sure that your work area is clean and free of any materials not needed for this activity. If you are allergic to lemons or other citrus fruits, juice drinks, or sugar, or if you have any other health issue, such as diabetes, that limits what you can eat, tell your teacher and do not taste the drink samples in this activity.

PROCEDURE

1. Record your group number (found on the sample cups) in your science notebook; this represents the batch of medicine you received.

2. Fill your tasting cup half-full of yellow lemon drink by carefully pouring from the sample cup into your tasting cup.

3. Taste the yellow lemon drink. Empty the cup.

4. Fill your tasting cup half-full of pink lemon drink.

5. Taste the pink lemon drink.

6. *Did the pink lemon drink taste the same, better, or worse than the yellow lemon drink?* Record your response in a table like Table 1 on the next page.

➢

Table 1: Results of Treatment

	Same as yellow lemon drink	Better than yellow lemon drink	Worse than yellow lemon drink
My response (Show with an X)			
My group's response (Show number of each)			

7. Share your results with your group. Summarize your group results in Row 2 of your data table.

8. Have one person from your group report your group's results to your teacher.

9. After a class discussion of the results, record the class's results and create a bar graph of the class's data on Student Sheet 3.1, "Analysis of Clinical Trial."

ANALYSIS

1. What evidence do you have that the medicine does or does not work to improve headaches?

2. **a.** What is a placebo?

   b. Why is a placebo group included in clinical trials?

3. In this activity, if a person finds that the drink tastes worse, the headache is gone, but there are side effects.

a. Assume that the side effects are mild, such as a slight stomachache. Explain why this medicine should or should not be sold to people suffering from a headache. Are there any trade-offs involved in your decision?

b. What if the side effects were serious, such as nausea and vomiting? Explain why this medicine should or should not be sold to people suffering from a headache. Are there any trade-offs involved in your decision?

4. In this activity, if a person finds that the medicine tastes better or worse, the headache is gone. Review the results of this simulation. Think about whether the medicine works and how often side effects occur. What would you conclude about the safety and effectiveness of this medicine for treating headaches? Support your conclusion with evidence.

EXTENSION

Go to the SALI page of the SEPUP website to find out how to post your class results on the site. Look at the results posted by other students. How do your results compare?

READING

How can we be certain that the foods we eat are safe, that the cosmetics we use won't harm us, and that medical products are effective? In the United States, the Federal Food, Drug, and Cosmetic Act and several other laws protect the public's safety. Before the Federal Food, Drug, and Cosmetic Act was passed in 1938, most products were not regulated.

CHALLENGE

How are medicines, such as over-the-counter and prescription drugs, tested?

READING

The Role of the FDA

Today, the U.S. Food and Drug Administration (FDA) enforces the laws on product safety and effectiveness. The FDA approves medicines that have been proven to be safe and effective for people to use. To be approved, a product must undergo and pass scientific tests. Usually, new medicines are first tested in animals to see whether they cause any harmful side effects. Only then are the medicines tested by volunteers in clinical trials. The results of clinical trials must show that the medicine is both safe and effective before it can be approved for use by the public.

What are two reasons that medicines must be tested before they are made available to the public?

Informed Consent

In clinical trials, volunteers are chosen carefully. They must not be allergic to drugs similar to the one that will be tested. Also, they cannot be taking certain other medications. These things might make it more likely for them to experience harmful side effects. The volunteers must be told about any risks, such as possible side effects. They sign an **informed consent** form. This form states that they have been told (informed) about the risks and that they agree (consent) to participate in the trial.

STOPPING TO THINK 2

a. Why do volunteers in a clinical trial have to sign informed consent forms?

b. Make a list of all of the information that you think should be included in an informed consent form.

Treatment and Control Groups

The volunteers are then divided into two groups. One group receives a pill or liquid that contains the medicine being tested. The other group receives a pill or liquid that looks the same but contains only an inactive ingredient, such as sugar. This inactive pill is called a **placebo** (pla-SEE-bo). Neither the patients nor their doctors know who is receiving the medicine and who is receiving the placebo. The placebo is used as a **control**. It helps to prove that any improvement in the patient is due to the medicine and not to other aspects of the medical treatment. For example, just going to see a doctor may result in an improvement in a patient's condition.

STOPPING TO THINK 3

a. Explain the relationship between a placebo and a control.

b. Why do you think the doctor isn't told whether the patient is receiving the treatment or the placebo? Imagine you are the doctor. How would you act if you thought your patient was getting a treatment that will work? How would you act if you thought your patient was getting just a placebo? How might this affect the patient?

c. What is the purpose of a placebo?

The Placebo Effect

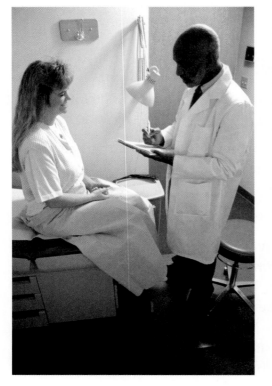

Why is it necessary to use placebos in clinical trials? For many years, scientists observed that people in clinical trials who receive placebos were more likely to show an improvement than other patients with the same disease. This improvement in the health of people receiving placebos is known as the **placebo effect**. Scientists give several explanations for the placebo effect. One factor is the regular, and often better, medical care that patients receive in a clinical trial. Another factor is the psychological effect of participating in a clinical trial. A patient who is getting regular medical attention and a pill may have a more positive attitude toward his or her health. He or she may also be more careful about diet, exercise, or other factors related to the illness. As a result, this person may be more likely to recover because of this positive attitude, and not as a result of anything in the pill. Today, scientists and doctors are studying this connection between a person's mental attitude and his or her physical health to understand how they influence each other.

STOPPING TO THINK 4

In your own words, what is the placebo effect?

The placebo effect sometimes makes it difficult to determine if a treatment is effective. For example, the use of vitamin C to treat colds is controversial. Some studies suggest that vitamin C improves cold symptoms. Other studies suggest that vitamin C doesn't work any better than a placebo, and that people who feel better from vitamin C are experiencing a placebo effect. Studies of high blood pressure, asthma, pain, depression, and cough have shown that about 30–40% of patients taking a placebo experience some relief of their symptoms. This relief can sometimes be measured objectively: for example, blood pressure actually drops in some patients taking placebos.

ANALYSIS

1. In clinical trials of medicines, why is one group of volunteers given a placebo? Explain.

2. Activity 3, "Testing Medicines: A Clinical Trial," simulated a clinical trial that investigated whether a headache medicine was effective. A person reporting a better or worse taste with the pink lemon drink (compared with the yellow lemon drink) represented a person feeling better after taking a pill for headache relief. In this simulation:

    a. What represented the medicine?

    b. What represented the placebo?

    c. Look at your data on Student Sheet 3.1, "Analysis of Clinical Trial." How many people experienced the placebo effect?

    d. Look at your data on Student Sheet 3.1. How many people in the placebo group were unaffected by the placebo?

3. Imagine a clinical trial to test a treatment for serious illnesses, such as heart disease or cancer. What is the trade-off of giving placebos to some people participating in this clinical trial?

Table 1: Clinical Trial of Cold Medicine				
	Feel the Same	Feel Better	Feel Worse	Total Number of People in Group
Control Group (Received placebo)	60	35	5	100
Treatment Group (Received medicine)	10	80	10	100

4. Review the data shown above from a clinical trial of a cold medicine.

a. Copy and complete a table like the one shown below to compare the number of people who feel better as a result of the medicine vs. the placebo:

Table 2: Analysis of Clinical Trial Data

	Number of people who feel better	Total number of people in group	Percent who feel better
Control group (Received placebo)			
Treatment group (Received medicine)			

b. Is the medicine effective in a high percentage of the population? Explain your answer.

c. Compare the percentage of people who feel worse in the placebo group vs. the percentage who feel worse in the treatment group. What difference do you observe? What could explain this difference?

d. Would you conclude that this medicine is safe and effective for treating colds? Explain the evidence for your conclusion.

Table 3: Clinical Trial of Headache Pill				
	Feel the Same	Feel Better	Feel Worse	Total Number of People in Group
Control Group (Received placebo)	50	46	4	100
Treatment Group (Received medicine)	41	55	4	100

5. Review the data shown above from a clinical trial of a headache pill. Think about whether the medicine works and how often side effects occur. Explain why this medicine should or should not be sold. Support your answer with evidence about the safety and effectiveness of this medicine.

6. a. Reflection: What evidence do you use to decide if cold, headache, or other over-the-counter medicines work for you?

b. What effect do you think a positive attitude has on your everyday health? What about when you are sick?

Good scientific experiments are often designed to test only a single factor, or **variable**. You may remember that Dr. Goldberger had screens put on the windows during his experiment on prisoners. He also had bedsheets and clothes washed regularly. One strength of Dr. Goldberger's experiment was his effort to reduce all the other variables—such as the presence of insects or variations in cleanliness—that could affect his results. His goal was to make diet the *only* factor that was being changed. In this case, diet was the variable being tested.

Scientists are interested in how people respond to the environment. People use their senses—touch, sight, hearing, smell, and taste—to get information about their surroundings. This information then travels through nerves to the human brain. In the next two activities you will investigate your sense of touch. Can you identify all of the variables that might affect your results? What can you do to try to keep all of these variables the same?

CHALLENGE

Which part of your arm is most sensitive to touch?

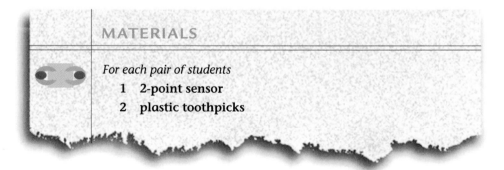

MATERIALS

For each pair of students
1 2-point sensor
2 plastic toothpicks

SAFETY

Be careful when doing the touch tests. Press gently when testing, making sure to only slightly depress the skin surface.

PROCEDURE

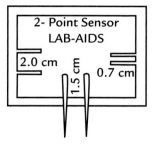

1. Slide 2 plastic toothpicks into the 2-point sensor on the side marked "1.5 cm."

2. With your eyes open, investigate your sense of touch by touching the skin of your fingers, palm of your hand, and forearm with the point of just *one* toothpick.

3. With your eyes open, touch your fingers, palm, and forearm with the points of both toothpicks.

4. Record your observations in your science notebook while your partner investigates his or her own sense of touch.

5. Have your partner close his or her eyes while you touch the skin on his or her fingers with either one or two toothpick points. Touch just hard enough to see that the points are barely pushing down on the skin. Randomly alternate between one and two points. Can your partner tell the difference?

6. Create a larger version of the table shown on the next page. In the table, record your observations about your partner's ability to tell the difference between one and two points on his or her fingers.

➢

Table 1: Observations of Touch Sensitivity

Person Being Tested	Fingers	Palm	Forearm
(Name)			
(Name)			

7. Repeat Steps 5 and 6 on your partner's palm *and* forearm.

8. Switch places and repeat Steps 5–7.

9. In your group of four, use Analysis Questions 1 and 2 to discuss your results.

ANALYSIS

1. Which part of your arm—your fingers, palm, or forearm—was the most sensitive to touch? What data do you have to support your conclusion?

2. In your group, how many people found fingers to be the most sensitive part of their arm? How many found palms or forearms to be the most sensitive? How similar were different individuals' responses to touch?

3. Why was it important for the person being tested to close his or her eyes?

4. Before scientists make comparisons, it is important that they perform well-designed experiments. In a well-designed experiment, all of the variables, except the one being tested, are kept the same.

  a. In your experiment, what variables did you keep the same?

  b. Were there any variables (except for the one being tested) that you could not keep the same?

READING

You probably found that some parts of your arm were more sensitive than others. In this reading you will learn about some of the reasons for these differences.

CHALLENGE

Why do different parts of your body have different sensitivities to touch?

READING

You use your senses, to gather information about your environment. Your senses include sight, hearing, taste, smell, and touch. To understand how you feel objects you touch, you need to know a little about your nervous system. Your **nervous system** includes your brain, your spinal cord, and your nerves. Figure 1 shows how these parts of your nervous system are connected. Nerves are found throughout your body, from the tips of your toes to the top of your head.

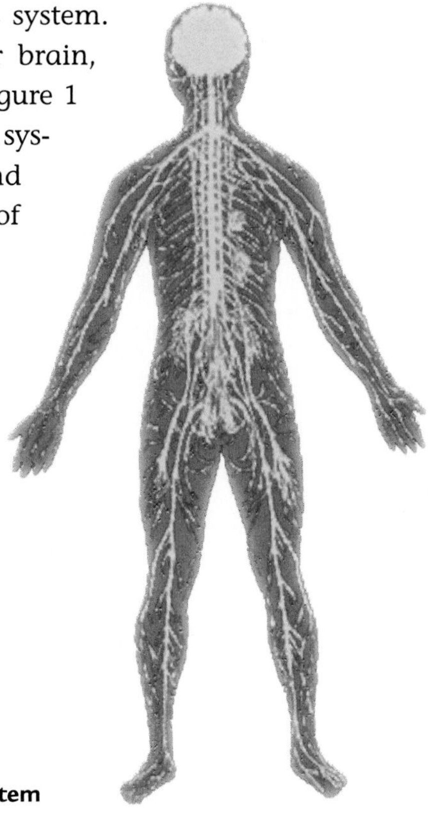

Figure 1: Human Nervous System

Your nerves take in information from the world. For example, your nerves have helped you collect information about the weight of your backpack when you pick it up and think, "It's heavy!" Information travels from nerves all over your body to your spinal cord before continuing to your brain. Your brain helps you understand what your nerves have detected. Your brain can then provide directions to other nerves. These other nerves send signals to your muscles that cause you to move.

Figure 2: Responding to the Environment
Signals travel from nerves in your body to your brain. These signals can travel up to several hundred miles per hour!

......

STOPPING TO THINK 1

a. Someone accidentally bumps into you in the hallway. What part of your body detected the feeling of being bumped?

b. Where in your body is this feeling analyzed?

......

Sometimes your nervous system responds to messages that you provide. For example, you decide to eat a sandwich. Your brain sends signals along nerves to direct your muscles to pick up the sandwich and take a bite. But what happens after you swallow? Do you have to think about directing your body to process the food? Your body automatically moves food through your digestive system and processes it. For this to happen, muscles inside your body must move. These muscles are controlled by the automatic, or involuntary, part of your nervous system. You don't even have to think about it!

......

STOPPING TO THINK 2

Is breathing completely involuntary? Explain.

......

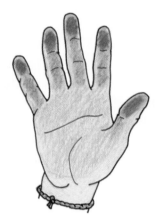

Figure 3: Touch Receptors on Human Hand

Why are some parts of your body more sensitive to two points than others? When something touches your arm, you feel it if it is detected by the nerve endings in your skin. You have nerves in your body that detect pressure, heat, sounds, smells, and light. The nerve endings that detect pressure on your skin are called **touch receptors** (ree-SEP-tors). They help carry a message from your skin to your brain.

Some parts of your body have more touch receptors than others. When two points stimulate the same touch receptors, you feel the points as one touch. When they stimulate different touch receptors, you feel two different touches. Figure 3 shows the concentration of touch receptors on a hand. Notice that the tip of the finger has more touch receptors than the rest of the finger. When you reach out to touch something, you often use your fingertips. You may have heard of people with limited or no vision reading Braille. Braille is a written language that uses raised dots instead of letters. Braille is read with fingertips. Not surprisingly, you have the greatest number of touch receptors right at your fingertips—just where they are needed.

EXTENSION

For links to more information on the human nervous system, go to the SALI page of the SEPUP website.

ANALYSIS

1. **a.** Where would you expect to have more touch receptors: on the palm of your hand or on the back of your hand? Explain your ideas.

   b. Explain how you could test your answer to Question 1a.

2. Review your results from Activity 5, "Can You Feel the Difference?" Based on what you now know, where on your arm—fingers, palm, or forearm—do you have the fewest touch receptors?

In the last two activities, you learned about your sensitivity to touch on different parts of your arm. How can you find out if what you learned about yourself applies to other people? You and your classmates will collect and compare data to find out how sensitive each of you are to touch on the palm of your hand. Do you think everyone has the same sensitivity to touch?

CHALLENGE

What is the smallest distance at which you can still feel two points? How does this compare with other people in your class?

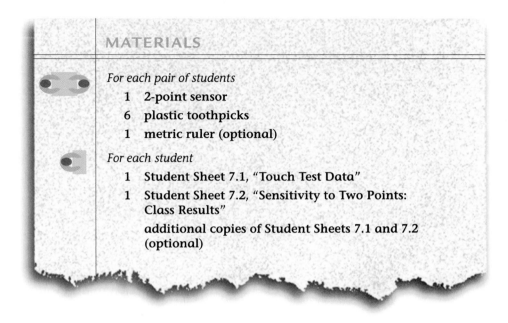

MATERIALS

For each pair of students

1   2-point sensor
6   plastic toothpicks
1   metric ruler (optional)

For each student

1   Student Sheet 7.1, "Touch Test Data"
1   Student Sheet 7.2, "Sensitivity to Two Points: Class Results"
    additional copies of Student Sheets 7.1 and 7.2 (optional)

SAFETY

Be careful when doing the touch tests. Press gently when testing, making sure to only slightly depress the skin surface.

PROCEDURE

1. *What is the smallest distance—0.7 cm, 1.5 cm, or 2.0 cm—at which you think you can still feel two points on the palm of your hand?* In your science notebook, record your hypothesis. Be sure to explain *why* you made this prediction.

2. Identify your dominant hand. (This is usually the hand you write with.) Throughout the experiment, you will test your dominant hand.

3. Begin completing Student Sheet 7.1, "Touch Test Data." Write in your name and circle which of your hands is dominant. You will begin by testing the palm of your hand, so circle "palm" as the part of the arm being tested.

4. Since you will test your partner (and vice versa), switch student sheets so you can record his or her data on his or her sheet.

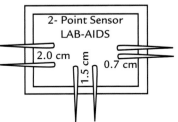

5. Slide 2 toothpicks into each side of the 2-point sensor as shown on the left. You should end up with toothpicks on three sides, with the toothpick points 0.7 cm apart, 1.5 cm apart, and 2.0 cm apart.

➤

6. Practice using the 2-point sensor so that you can safely and easily test using any of the three sides.

7. As the experimenter, you will use the 2-point sensor to test your partner. Record your partner's responses on Table 1, "Touch Response," on Student Sheet 7.1. *It is important that you move across each row in Table 1 as you record your partner's response.*

   For example, when you do Trial 1, you will work across the first row:

   a. Turn the 2-point sensor to the 0.7 cm side and touch your partner's palm with just one point.

   b. Turn the sensor to the 1.5 cm side and touch your partner's palm with just one point.

   c. Turn the sensor to 2.0 cm and touch your partner's palm with two points.

8. Before starting the touch tests, ask your partner to close his or her eyes. *The partner being tested should not try to "guess the right answer." The goal is to report what you really feel—one point or two.*

9. Use Table 1, "Touch Response" to test your partner. *You can start with any row you want, but be sure you complete all the rows. Don't tell your partner which row you are using.* Remember to touch just hard enough to see that the points are barely pushing down on the skin. After each touch test, have your partner report whether he or she feels one or two points and record the response.

10. After you complete all the touch tests in Table 1, have your partner test you by repeating Steps 7–9.

11. Give your partner back his or her original Student Sheet.

12. Complete the rest of Student Sheet 7.1.

13. If you have time, repeat this experiment to test the sensitivity of your fingertips and your forearm.

ANALYSIS

1. **a.** According to your data, what can you conclude about your sensitivity to 2-point touches? How does this conclusion compare with your hypothesis?

   b. Compare your results with those of your partner. How similar or different are your results?

   c. Compare your results with those of another pair of students. How similar or different are your results?

2. Look at the class results on Student Sheet 7.2, "Sensitivity to Two Points: Class Results." Compare the smallest distance at which you could feel two points with the results of the rest of the class. What can you conclude about the sensitivity of different people to touch? Is it possible to make conclusions about people in general?

3. You were able to determine the smallest distance at which you could still feel two points using only the 2-point touch data. The 1-point touches acted as a control. Why would you need a control when experimenting on people?

4. **a.** A good experiment is reproducible. What parts of this experiment are reproducible?

   b. How could this experiment be improved?

5. What factors make studying people scientifically difficult? How do scientists deal with these factors in a well-designed investigation?

6. Design an experiment to determine a person's sensitivity to sound. Assume you have a machine that you can set to produce sounds of varying volume (measured in units called decibels). **Hint:** Think about how you tested your sensitivity to 2-point touches.

EXTENSION

Go to the SALI page of the SEPUP website to find out how to post your class data on the site. Look at the data posted by other students. What can you conclude about the sensitivity of different people to touch? What effect does sample size have on your conclusions? Explain your ideas.

READING

Studying people *scientifically* presents some interesting challenges. One challenge is that different people react differently to identical events. Think about the results of your experiments on touch sensitivity. You probably discovered that not everyone could feel two points at the same distance. How do you think scientists deal with this type of challenge?

CHALLENGE

How are qualitative and quantitative data used when testing a hypothesis about people?

READING

Scientists usually begin with an idea about what they want to investigate. This is also true for scientists who study people. This idea, or hypothesis, is the basis for their study. As you now know, a **hypothesis** (hi-PAH-thuh-sis) is an explanation based on observed facts or on an idea of how things work. For example, Dr. Goldberger hypothesized that there was a relationship between pellagra and diet.

Doctor and patient

In some cases, scientists predict the possible outcomes of an experiment. A hypothesis can include a prediction of what might happen. In the last activity, your hypothesis included a prediction about the results. To make sure they don't influence results, scientists studying people often do *not* predict the possible results of their experiments.

Scientist examining rat brain scan

After developing a hypothesis, scientists usually plan and conduct experiments. The results of an experiment can provide evidence for or against a hypothesis. In some cases, an experiment provides new information. This new information can produce another hypothesis and another experiment. In this way, a hypothesis can be a "work in progress" that is continually revised.

Doctor and nurse working on computer

*The word **data** (DAY-tuh) is plural. It refers to a set of information. The singular form is* datum, *one piece of information.*

STOPPING TO THINK 1

You hear your friend Yoshi tell someone that a hypothesis is the same as a guess. Explain whether you agree with Yoshi.

In the case of pellagra, Dr. Goldberger conducted an experiment on prisoners to test his hypothesis. During the experiment, 7 out of 11 prisoners, or 64%, developed pellagra after six months. Even though all of the prisoners ate the same diet, they did not respond in the same way. Scientists try to account for individual differences by studying a large enough **sample size**.

STOPPING TO THINK 2

Explain whether Dr. Goldberger provided enough evidence to prove that pellagra is not contagious.

In the United States today, government agencies, such as the Food and Drug Administration (FDA), tightly control and review experimentation on humans. After receiving government approval, companies often search for volunteers to participate in clinical trials. Clinical trials test medicines, food, medical procedures, and medical equipment. **Data** collected during these trials provide evidence for making conclusions about a product or treatment. The type of data collected depends on what is being investigated.

For example, imagine a company that is conducting trials of a new hearing aid. The company would like to know how clearly volunteers can hear when using the hearing aid. People will try out the hearing aid and report how clearly they hear sounds, and whether they hear any buzzing or other undesired noise from the hearing aid. They may also report whether they find the hearing aid comfortable to wear and attractive in appearance. These are all examples of qualitative data. **Qualitative** (KWAL-i-tay-tiv) **data** describe properties or characteristics (qualities) that are used to identify things. Qualitative data often result from putting things into categories.

➤

STOPPING TO THINK 3

You notice a skateboard for sale on an Internet website. The price is good, but the website provides no picture or other information. What qualitative data would you like to have before you decide if you will buy it?

What if you were interested in buying the hearing aid? You would probably want to know how well it works when compared with other hearing aids. In clinical trials, people with hearing problems might be tested to measure the softest sound they could hear with and without the different hearing aids. The volume of sound (how loud it is) is measured in decibels. The company might report that in clinical tests 90% of the people wearing the hearing aid heard sounds as low as 35 decibels, when only 10% could hear these sounds without the hearing aid. In this case, numerical, or quantitative, data were collected. **Quantitative** (KWAN-ti-tay-tiv) **data** are values that have been measured or counted. The word *quantitative* is related to the word *quantity*, which means "number" or "amount."

Consider an everyday example of qualitative and quantitative information. Imagine you are at a large store with a friend when suddenly you turn around and she's not with you anymore. You look around, but you don't see her anywhere nearby. You ask a store employee if he has seen her. You describe your friend as a tall 13-year-old girl with short brown hair, brown eyes, a red jacket, two braids, and a dimple when she smiles. You have included important qualitative and quantitative information about your friend.

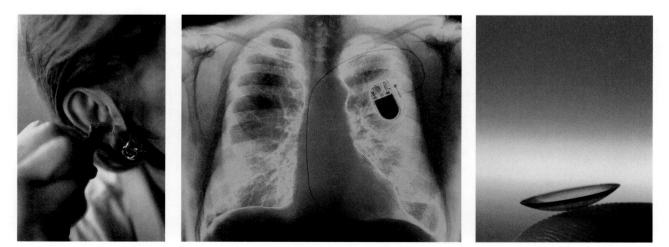

Medical products, such as hearing aids, pacemakers, and contact lenses are tested before being sold to the public.

STOPPING TO THINK 4

Reread the description of your friend in the above paragraph. Create a list of each of her characteristics. Identify each characteristic as either qualitative or quantitative.

Both qualitative and quantitative evidence are important in identifying your friend. Qualitative data provide information about important characteristics that are difficult to measure but can be described and categorized. But in some cases quantitative data help give a clearer description. Think about the description of your friend in the previous paragraph. A height that seems tall to you might seem short to the store employee. But if you tell him that your friend is 5 feet 7 inches tall, you will both have a clear idea of how tall she is.

STOPPING TO THINK 5

a. What kinds of qualitative data are useful in studying people scientifically? Provide at least two examples.

b. What kinds of quantitative data are useful in studying people scientifically? Provide at least two examples.

ANALYSIS

1. You decide to take a medicine for your upset stomach. You have a choice of two medicines. Both medicines are advertised as safe based on clinical trials. Medicine A was tested on 100 people. Medicine B was tested on 10,000 people.

   a. Which medicine would you take? Explain. Support your answer with evidence.

   b. Was your decision based on qualitative or quantitative information?

2. You're a volunteer at a local hospital. While you are there, you read a patient chart containing the data below. Identify each item of patient data as quantitative or qualitative.

3. Imagine conducting a clinical trial of a headache medicine. Based on your research, you hypothesize that the medicine will successfully treat headaches in people. Before the Food and Drug Administration (FDA) will approve your study, you must explain the type of data you will collect to test your hypothesis.

Skin tone: Flushed

Heart rate: 77 beats per minute

Weight: 129 pounds

Height: 5 feet 5 inches

Body temperature: 100°F (38°C)

Response to touch: Slow

   a. List at least three kinds of qualitative data you will collect. **Hint:** Think about what information you would collect from the volunteers and what information you would collect about the medicine in order to determine the safety and effectiveness of the medicine.

   b. List at least three kinds of quantitative data you will collect. **Hint:** Think about what information you would collect from the volunteers and what information you would collect about the medicine in order to determine the safety and effectiveness of the medicine.

4. Think about the activities you have done so far in this unit. What are the common elements of a well-designed experiment?

5. **Reflection:** Both qualitative and quantitative data provide evidence for making decisions. How have you used each of these types of data to make decisions? Describe your experiences.

LABORATORY

Because of differences among individuals, a **range**, or set of values, is often used to generalize to a larger group of people. For example, you may have had nurses take your pulse. What can your pulse tell them? Through scientific study, a normal pulse range for people at rest has been established. The normal range for adults is 60–80 beats per minute. The use of a range (60–80) instead of a specific number helps take into account the differences among people.

To establish a range, data are collected from many trials. Increasing the number of people tested increases the chance that the conclusions are true for most people.

If you play baseball, softball, or other games that involve throwing and catching, your teacher or coach may tell you to catch the ball with two hands. Does catching with two hands (as compared to one) really increase your ability to catch a ball?

CHALLENGE

How can you study people scientifically? Find out by collecting data on people and then designing your own experiment.

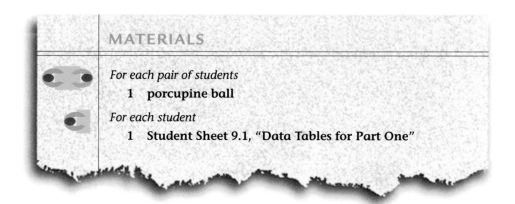

MATERIALS

For each pair of students
1   porcupine ball

For each student
1   Student Sheet 9.1, "Data Tables for Part One"

PROCEDURE

Part One: One vs. Two Hands

1. Based on what you know about catching balls, record a hypothesis about whether the ability to catch a ball increases when using two hands (vs. one).

2. With your partner, brainstorm all of the variables that you will try to keep the same while conducting the experiment.

3. Stand 2–3 meters away from your partner, as directed by your teacher.

4. Have your partner toss you the ball while you catch it using only one hand. Remember to keep all variables, except for the one being tested, the same.

5. *Did you find catching the ball with one hand easy, difficult, or somewhere in between?* Record your response in Table 1, "My Data," of Student Sheet 9.1, "Data Tables for Part One."

6. Have your partner toss you the ball 20 times while you continue to catch it using only *one hand*. Record the results of each catch in Table 2, "My Data: Number of Completed Catches." Place an "X" in the box when you catch the ball using one hand, and an "O" when you do not. When you are done, count and record the total number of completed catches.

7. Repeat Steps 3–6, but use *two hands* to catch the ball.

8. Now repeat Steps 3–7 for your partner to collect his or her own data.

9. Share your results with the class. Use the class data to complete Tables 3 and 4 on Student Sheet 9.1.

10. Complete Analysis Questions 1–4.

Part Two: Designing Your Own Experiment

11. Design an experiment to investigate the effect of one variable on your ability to catch a ball. For example, you may want to investigate your ability to catch with one eye closed, to compare your ability to catch with your right or your left hand, or to vary the size of the ball. When designing your experiment, think about the following questions:

   • What is the purpose of your experiment?

   • What variable are you testing?

   • What is your hypothesis?

   • What variables will you keep the same?

   • How many trials will you conduct?

   • Will you collect qualitative and/or quantitative data? How will these data help you to reach a conclusion?

   • How will you record these data?

12. Record your hypothesis and your planned experimental procedure in your science notebook.

13. Make a data table that has space for all the data you need to record. You will fill it in during your experiment.

14. Obtain your teacher's approval of your experiment.

15. Conduct your experiment and record your results.

16. If you have time, switch procedures with another pair of students. Conduct their experiment and record the results while they do the same with yours. Then exchange results so that each pair has two sets of results for the experiment they designed.

EXTENSION

Conduct your experiment with different age groups, such as children vs. adults. Compare your data with the data collected in class. Are your conclusions the same?

ANALYSIS

Part One: One vs. Two Hands

 1. Explain whether your own data supported or disproved your hypothesis about the ability to catch a ball with two hands (as compared to one).

 2. Look at the class data on Student Sheet 9.1.

   a. In Table 4, what was the range of students' ability to catch a ball with one hand?

   b. In Table 4, what was the range of students' ability to catch a ball with two hands?

   c. Use the class data to explain whether the ability to catch a ball increases when using two hands (as compared to one).

3. **a.** What qualitative data did you collect in Part One of this activity?

   b. What quantitative data did you collect in Part One of this activity?

   c. Which type of data was more useful for comparing results among the class members?

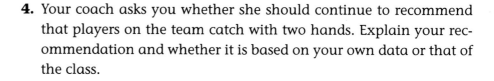 4. Your coach asks you whether she should continue to recommend that players on the team catch with two hands. Explain your recommendation and whether it is based on your own data or that of the class.

Part Two: Designing Your Own Experiment

5. What conclusions can you make based on the results of your experiment? Explain how your conclusions are based on the data collected during your experiment, and whether your hypothesis was supported or disproved.

6. a. In your experiment, what variables did you keep the same?

   b. Were there any variables (except for the one being tested) that you could not keep the same?

   c. How could you improve the design of your experiment? Explain.

7. Was it possible for another pair to repeat your experiment? Were the results similar to your original results? Explain.

8. How do scientists use good experimental design to collect reproducible data about people?

Conducting studies on people is time-consuming and expensive. Researchers need funding. The U.S. National Institutes of Health (NIH) is one organization that provides money for such studies. What does NIH look for when researchers request money to study people scientifically?

THE NIH COMMITTEE

Listed on the following pages are proposals for five studies requesting money to conduct clinical trials. Most of the studies are based on actual research projects. As a member of the NIH review committee, your job is to evaluate the experimental design of each proposed study. Based on your review, make a recommendation as to whether NIH should fund each study.

CHALLENGE

Which proposals have an experimental design worth funding?

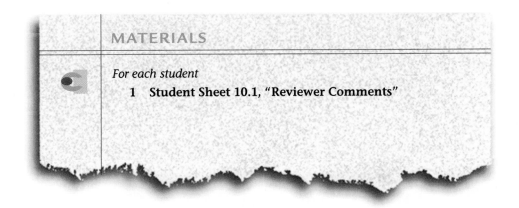

MATERIALS

For each student
1   **Student Sheet 10.1, "Reviewer Comments"**

PROCEDURE

1. With your group, brainstorm all of the factors that are important in designing an experiment to study people scientifically. Work together to agree on a complete list.

   • Listen to and consider the explanations and ideas of other members of your group.

   • If you disagree about a factor with others in your group, explain to the rest of the group why you disagree.

2. Read one of the studies. (The studies begin on the next page).

3. Use your list of experimental design factors from Step 1 and Student Sheet 10.1, "Reviewer Comments," to evaluate the study. Record the name of the study, the factors of its experimental design, and any other comments you have.

4. Based on your comments, decide whether you would recommend funding for this study.

5. Repeat Steps 2–4 for the remaining studies.

Study 1:
Clinical Trial of "Summer Fever" Medicine

We are working to prevent "summer fever," a contagious disease that kills or paralyzes many children each summer. This disease is spreading quickly across the United States. We have developed a medicine that we believe will prevent children from catching this disease. We would like to conduct a clinical trial to find out if our medicine is effective in children. Our plan is to give the medicine to 12,000 children in the state's capital city. We will count how many of these children catch summer fever. We are sure that all of these children will be completely protected as a result of taking this medicine.

To prove that the medicine is safe for humans, my assistant, my 14-year-old child, and I have all taken it. We have not suffered any bad effects from the medicine.

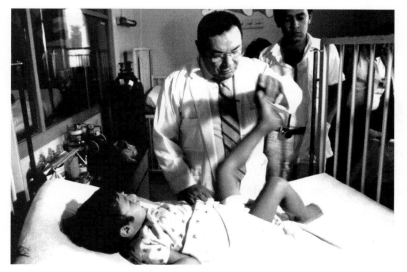

This child's legs are paralyzed as a result of a disease similar to "summer fever."

Study 2:
Clinical Trial of Burn Cream

We have developed a new, medicated cream that helps heal burns on animals. Faster healing reduces pain and reduces the chance that the burns will become infected. We would like to test this cream on humans to see if it helps heal burns faster. We plan to conduct clinical trials on 200 people ages 20–40. Half of the people will be men and half will be women. Each person must have third-degree burns that are no larger than 3 centimeters by 3 centimeters. They must also be healthy except for their burns.

Fifty of the men and fifty of the women will be treated with our new medicated cream. The other half will receive a placebo cream that does not contain the medication. We will not let either group know which cream they are receiving. Each person will apply 2 grams of cream to the burn every morning for 20 days.

Three studies done by the local veterinary school have shown that the medicated cream is effective in healing burns on rabbits, cats, and dogs. Each study involved 100 animals. Half of the animals received medicated cream and half received cream without medication. The burns on animals treated with medicated cream healed faster.

Study 3:
Clinical Trial of Weight Loss Method

More than half of the adults in the United States are overweight. Many of them try risky fad diets to lose weight. One way to help people lose weight is to find effective ways in which they can reduce the number of calories they eat. For this reason, we would like to conduct a clinical trial of one method of reducing calorie intake. We would like to find out whether drinking water with an appetizer before a meal will help people reduce the number of calories they eat.

We will begin with 24 volunteers who will be told about all aspects of the trial. We will select volunteers who are men between the ages of 35–55 and who are at least 20% overweight. In our control, each volunteer will eat a chicken-rice appetizer that contains exactly 200 calories. We will then count how many calories the volunteer eats for lunch after eating this appetizer. To measure the effect of drinking water with the appetizer, the same volunteer will eat the same appetizer on a different day, but this time with a 12-ounce glass of water. We will again count the number of calories the volunteer eats for lunch afterward.

Each volunteer will repeat the control and the experiment three times. We will then average the results to determine any change in the number of calories eaten in each part of the trial.

Study 4:
Clinical Trial of Relaxin

We have developed a new drug called Relaxin. We believe Relaxin can be used to calm and relax people as well as prevent nausea. For these reasons, we think this drug would be useful to pregnant women. We would like to conduct a clinical trial to find out if this drug is effective for pregnant women. We will provide Relaxin to some doctors. These doctors will be asked to prescribe Relaxin to pregnant

women who ask for a drug to prevent nausea. Of course, we will tell the women that they are getting Relaxin, because people testing new drugs should give their permission. We will then record the effects on these women by asking them if they feel calmer and less nauseated.

Relaxin has been tested for its effectiveness and safety. Our scientists conducted an experiment on mice. Half of these mice were given Relaxin, while the other half were given a placebo. We observed and recorded how much the mice moved. The mice that were given Relaxin moved less than the mice that were given the placebo. This showed that the drug is effective. None of the mice had any side effects from this drug. To further test the safety of Relaxin, we gave large doses to some mice. None of these mice showed any permanent side effects from the large doses.

Study 5:
Clinical Trial of Insulin

Today, insulin is used to treat diabetes. How did scientists discover how to control the effects of this disease? Watch a segment from the video, *A Science Odyssey*: "Matters of Life and Death," to find out. Evaluate the quality of the experiments shown in this segment. What other comments do you have about these experiments?

ANALYSIS

1. Which study—1 through 4—had the best experimental design? Explain.

2. Discuss your funding recommendations with your group. Do you agree on which studies, if any, should be funded? What other concerns do you have?

3. You find out that NIH has only enough money to fund one study and plans to fund the best one. Explain which study you would fund. Support your answer with evidence and identify the trade-offs of your decision.

   Hint: To write a complete answer, first state your opinion. Provide two or more pieces of evidence that support your opinion. Then consider all sides of the issue and identify the trade-offs of your decision.

4. Choose one of the studies. Review your comments about its experimental design and think about how the study could be improved. Rewrite the study to include your recommendations for improving the quality of the experiment.

5. **Reflection:** Based on what you have learned in this unit, how do scientists solve problems?

EXTENSION

Bring in news articles that describe scientific studies involving people. Analyze the studies according to the procedures you used in this activity. What issues do scientists face when studying people scientifically?

Body Works

Unit B

Body Works

"Hey Dad, did you fill out all those forms for the race?" asked Maya.

"Yes, Maya," her father replied. "The race organizers wanted to make sure that we were both in good health, so I had to answer questions about our health and list medications that either you or I were taking."

"Why do they need so much information? It's only a 5K race," said Maya.

"They need to make sure that they can take care of all the people who run in the race," explained her father. "Even though it's been several years since my heart surgery, I still take medications for my heart. I have to be a lot more careful with my health than I was before. I'm glad that they'll have all of this information, just in case."

"I sure wish you had never had any heart problems," sighed Maya.

"Me too," replied her father. "I could have taken better care of my health when I was younger, but I didn't. While it might not have prevented my heart problems, it would have reduced the risk. I hope that, as you get older, you'll make better choices than I did."

• • •

What choices do you make about your health every day? Which of these decisions may affect your health in the future? What types of information could help you make better decisions about your body? Would you be willing to fund scientific research to answer these kinds of questions?

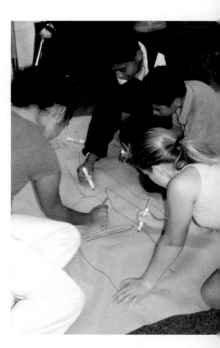

Answering such questions thoughtfully requires knowing and understanding scientific information about the human body. In this unit, you will learn more about your own body and how it works.

TALKING IT OVER

You may feel healthy most of the time. But when you feel sick, you may turn to medicines for relief. How do you decide whether to take a medicine, and what kind of medicine to take?

FEELING SICK

Sam and his older brother and sisters are talking one Saturday morning.

Sam: I feel so sick. It's just not fair! Yesterday I felt fine, but today I feel awful. I don't want to miss the birthday party at the twins' house tonight.

Andrea: You must feel pretty bad to talk about missing a party. What's wrong?

Sam: You don't really want to know. My stomach hurts so much it feels like an elephant sat on it. I had to get up and run to the bathroom four times last night and now I feel dizzy and weak.

Marta: Sounds pretty bad. Let me feel your forehead....Oh yeah, you are hot. You probably have a fever.

Andrea: Here's the thermometer. Let's take your temperature.

Robert: Don't we have something Sam can take to feel better?

Marta: Hmm, you have a fever of 101.2°F (38.4°C). Let me check the medicine cabinet.

Andrea: I say aspirin is the best answer. It always works for me when I get a headache.

Sam: It doesn't always work for me when I have a headache.

Robert: Uncle Richard takes an herbal remedy. I forget what he calls it. He says it works for him and it's safer than medicines.

Andrea: Yeah, but Aunt Jan says herbal remedies aren't tested the way medicines are. She says you can't be sure they work and if they're safe.

Marta: We have fever medicines, herbal tea, and some stomach medicine. I could also give you some ginger ale or soup. That's what Mom always has when she is sick.

Robert: Why don't we get them and look at the labels?

Marta: Good idea. And I'll call the doctor and describe your symptoms. Dad left the number by the phone.

CHALLENGE

What are the trade-offs of taking a medicine when you feel sick?

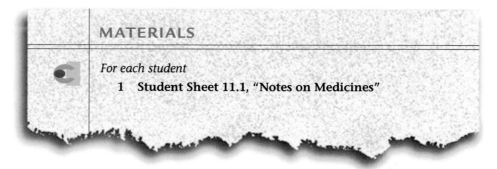

MATERIALS

For each student
1   **Student Sheet 11.1, "Notes on Medicines"**

PROCEDURE

1. Imagine you have the same symptoms Sam has. In your science notebook, record what you would do, or have done, in Sam's position.

2. Have each member of your group share what he or she wrote. Listen to each person, but do not comment.

3. Read the doctor's response to Marta to learn more about Sam's choices:

   "It sounds like Sam has a common stomach virus. He should feel better in a day or two as long as he drinks plenty of liquids. It's important that he doesn't get dehydrated. If his fever doesn't get any worse, Sam can decide for himself whether he would like to take any medicine.

   "Fever and diarrhea are sometimes ways that the body fights off bad germs. They're the body's way of trying to make you better. But a fever reducer or diarrhea medicine may help Sam feel better now. It's up to him. I'm leaving town for a couple of days, so call the hospital if his fever goes up or if he doesn't feel better in two days."

4. Shown below are the labels from the medicines that Robert found in the medicine cabinet. Use Student Sheet 11.1, "Notes on Medicines," to help you analyze the labels. Read them carefully as you decide what Sam should do.

Medicine A: Pain Reliever/Fever Reducer Tablet

Warning: If you consume three or more alcoholic drinks every day, ask your doctor whether you should take this product or other pain relievers/fever reducers. This product may cause liver damage. **Do not use** with other pain killers/fever reducers unless directed by a doctor. Keep this and all drugs out of the reach of children.

Stop using and consult a doctor if:
· symptoms do not improve · new symptoms occur · pain or fever persists or gets worse · redness or swelling is present

---

Medicine B: Pain Reliever/Fever Reducer Tablet

Warning: Children and teenagers should not use this medicine for chicken pox or flu symptoms before a doctor is consulted about Reye's Syndrome, a rare but serious illness associated with this medicine. Do not take this product if you are allergic to aspirin or if you have asthma, unless directed by a doctor.

Drug Interaction Precaution: Do not take this product if you are taking a prescription drug for anticoagulation (thinning the blood), diabetes, gout, or arthritis, unless directed by a doctor.

Medicine C: Diarrhea Tablet

Warning: Keep this and all drugs out of the reach of children. Do not use for more than two days unless directed by a physician. Do not use if diarrhea is accompanied by high fever (greater than 101°F), or if blood or mucus is present in the stool, or if you have had a rash or other allergic reaction to this medication. If you are taking antibiotics or have a history of liver disease, consult a physician before using this product.

Medicine D: Herbal Tea

This plant grows wild in the United States, and Native Americans have used it for centuries for a variety of purposes. This tea contains a blend of leaves, flowers, and stems of an organically grown plant. In modern studies, leaves, flowers, and roots of this plant have shown measured effects in supporting the immune system.*

*The Food and Drug Administration has not evaluated these statements. This product is not intended to diagnose, treat, cure, or prevent any disease.

ANALYSIS

1. What kinds of information are provided on the labels?

2. Sam decided to make a decision based on the information he has. Think about all of the options available to Sam. He could

    a. drink liquids and wait until he feels better,

    b. drink liquids and take one of the medicines, or

    c. drink liquids and take a combination of the medicines.

    If you were Sam, what would you do? Assume that Sam's medical history is the same as your own. Support your answer with evidence and identify the trade-offs of your decision.

    Hint: To write a complete answer, first state your opinion. Provide two or more pieces of evidence that support your opinion. Then consider all sides of the issue and identify the trade-offs of your decision.

3. Explain whether you used more qualitative or quantitative evidence to make your decision. Support your answer with examples.

4. What are the trade-offs of taking a medicine when you feel sick?

INVESTIGATION

Organs in the human body are divided into **systems** based on their function. For example, consider the excretory (ECK-skruh-tor-ee) system. The function of the excretory system is to remove liquid waste from the body. The kidneys help perform this function, so they are a part of the excretory system. You may be familiar with the organs and functions of other systems, such as the digestive or circulatory systems. Use your knowledge of the human body to look more closely inside yourself. In the photos below, which organ systems help each of the students do the activities shown?

CHALLENGE

What do you know about the human body and how it functions?

MATERIALS

Part One: Laying It Out

For each group of four students
- 1 sheet of chart paper
- 4 markers of assorted colors
- glue or tape (optional)

For each student
- 1 Student Sheet 12.1, "Functions of Human Body Systems"
- scissors (optional)

Part Two: Modeling the Human Body

For each group of four students
- 4 differently colored sticks of modeling clay
- 1 human body torso model

For each student
- Student Sheets 12.2a and 12.2b, "Human Body Systems"

PROCEDURE

Part One: Laying It Out

1. With your group, draw an outline of a human body on a piece of chart paper.

2. Have each person in your group take a marker. Work together to do the following:

   a. Draw in the major internal organs of the body.

   b. Write down what you think each organ does.

   c. Around your outline write questions you have about the human body.

3. Go around the room to view the drawings of other groups.

4. Use your own knowledge as well as information from the Introduction and Table 1 to work by yourself to complete Student Sheet 12.1, "Functions of Human Body Systems."

Part Two: Modeling the Human Body

5. Place the human body torso model on a flat surface.

6. As a group, use Student Sheets 12.2a and 12.2b, "Human Body Systems," and the modeling clay to make a 3-dimensional **model** of some of the organs and structures in the human body. In this model the plastic represents skin.

7. Use the modeling clay to create each of the organs listed in Table 1. Student Sheets 12.2a and 12.2b can help guide you in forming the organs correctly.

8. Place the organs into the bottom half of the human body torso model. Follow the order listed in Table 1 by placing the first structure (the muscles) down first and then adding the others in the order listed. Remember to use Student Sheets 12.2a and 12.2b for help. **Hint:** You'll need to put the end of the digestive system behind the bladder to make it accurate.

9. When you are done modeling the organs, place the other half of the plastic model on top of the body. You have now created a model of your internal organs.

10. Compare the placement of the internal organs in the model to your own body. Try to figure out where these organs are in your body.

11. Take your model apart. Roll the modeling clay back into separate balls of each color.

Table 1: Organs and Structures to Model	
Clay Color	**Organs and Structures**
	muscles of back and buttocks
	spinal cord
	kidneys (connected to bladder by thin tubes)
	esophagus
	stomach
	small intestine
	large intestine (and rectum)
	liver
	windpipe (trachea)
	lungs
	heart
	bladder
	rib cage (ribs and sternum)

ANALYSIS

1. The liver is the largest internal organ of the human body. Was the liver the largest organ in your model? Do you think that the other organs you modeled were accurate in size? Why or why not?

2. Was the model that you created in Part Two a good model of the human body? Why or why not?

3. Prepare a table with headings as shown below. Fill in the first column with the organs or structures listed in Table 1.

   a. In the second column of your table, identify the system that matches each organ or structure. For example, the stomach is a part of the digestive system.

   b. In the third column of your table, identify the function of each of the systems you mentioned in 3a.

Organs and Structures	System	Function

4. **Reflection:** What new things have you learned about the human body in this activity?

ROLE PLAY

How often do you think about what's going on inside your body? Most healthy people don't need to worry about what's happening inside. But knowing more about the human body can help you make better decisions about your health.

• • •

For links to more information about the liver, go to the SALI page of the SEPUP website.

CHALLENGE

How does the liver help your body stay in balance?

PROCEDURE

1. Assign a role for each person in your group. There are four roles: Mr. Lee, Rick, Kamika, and Yolanda. Assuming there are four people in your group, each of you will read one role.

2. Read the following role play aloud.

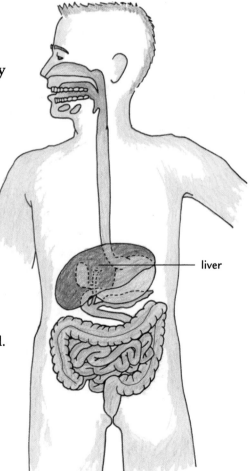

liver

LIVING WITH YOUR LIVER

Rick: Mr. Lee, you told me that we're going to study the digestive system soon. I think it's going to be pretty interesting.

Yolanda: I think digestion is gross. My mom's a surgeon and she showed me some pictures of the inside of…

Kamika: Relax, Yolanda. Rick cares about the digestive system because he almost died from something he ate this summer.

Mr. Lee: Almost died? What happened, Rick?

Rick: My little sister and I found some mushrooms growing near our house. I ate some. She doesn't like it when Dad puts mushrooms in our salads, so she didn't eat any.

Yolanda: My grandpa says you should never eat wild stuff whose name you don't know.

Rick: Well, I sure won't do it again. I never felt so awful. The nurse in the emergency room said I was lucky my dad brought me in so fast.

Kamika: So the mushrooms burned your stomach?

Mr. Lee: Actually, most poisons are dangerous because they can destroy your liver.

Yolanda: Why is that? I thought the liver was just one of those weird organs that doesn't really do much.

Rick: Nope. It does a whole lot of stuff your body can't live without.

Yolanda: Isn't the liver huge? Isn't it the size of your brain?

Mr. Lee: That's close, Yolanda. In fact, it's even bigger! A liver weighs over 3 pounds. It's the largest organ in your body, except for your skin.

Kamika: So Rick almost died because those mushrooms hurt his liver? That doesn't make sense. The food you swallow doesn't even go to the liver.

Mr. Lee: That's true, but after the food is broken down, your blood carries the substances you've digested to the liver first. The liver controls which substances get stored or filtered out. Only then are these substances carried to the rest of your body.

Rick: Oh, I get it. The liver's sort of like a traffic cop that regulates which cars go and which cars stop.

Mr. Lee: Exactly. That's a great metaphor, Rick!

Yolanda: I remember my mom's friend talking about this—she's a toxicologist. The liver has to break down harmful substances so they don't get to the rest of your body and hurt other organs.

Kamika: Now I get it. Rick's liver had so much toxic stuff sent to it all at once that it got damaged.

Mr. Lee: You're right, Kamika. Even now Rick probably has to be careful what he eats while his liver recovers.

Rick: Yeah. That's because the liver also helps you digest fats, and helps regulate how much cholesterol (kuh-LESS-tuh-rall) and sugar are in your blood.

Yolanda: It sounds like the liver controls how much and what kinds of substances go to different organs and systems.

Mr. Lee: Yes, the liver helps your body keep in balance. That's what **regulation** is—keeping things balanced and responding to changing needs. The liver does many things in your body, but most of them involve regulation.

Kamika: What's a healthy balance of a mushroom poison in your blood? Zero, I bet!

Rick: I think so!

Kamika: But maybe that's not true for cholesterol and sugar. Wouldn't you need some around all the time so your body can use it?

Yolanda: Yeah, my mom says you need some cholesterol. *Too much* cholesterol is the problem.

Mr. Lee: And sugar is what we use as a quick source of energy, but too much or too little in the blood can be a serious problem!

Rick: Mr. Lee, I overheard a doctor telling my parents that if I were an adult, he would tell me not to drink any alcohol while my liver was recovering.

Yolanda: Why would that be?

Kamika: Maybe it's harmful!

Mr. Lee: That's right, Kamika. It is harmful. We call these harmful substances toxins. A **toxin** (TOX-in) is any substance that can cause damage to your body. Alcohol is a toxin. It can cause a lot of damage if someone drinks a large amount all at once or smaller amounts over long periods of time. If someone's liver is already damaged, alcohol can be toxic in even smaller amounts.

Rick: So your liver can wear out, but a little bit at a time.

Mr. Lee: That sounds right, Rick. Sometimes damage to the liver builds up over many years as the liver works and works to remove toxins. Scar tissue forms, which is called cirrhosis (si-ROW-sis). If it's bad enough, you need a transplant.

Kamika: Cirrhosis! That's what my cousin has! I didn't realize it meant a worn-out liver. But she's only 19 and she doesn't touch alcohol. She has hepatitis (hep-uh-TIE-tus). She got it from a blood transfusion when she was a baby.

Mr. Lee: Your cousin must have hepatitis C. Until about 1990 there was no way to check donated blood to find out if it was infected. There's still no vaccine for hepatitis C, but there is for hepatitis A and hepatitis B.

Yolanda: Hepatitis attacks the liver. They made sure to vaccinate my uncle for hepatitis A and B as soon as they saw he had liver damage.

Rick: So Kamika's cousin won't ever be allowed to drink, I guess. Hey, you know what other things are toxins? Ibuprofen (eye-byoo-PRO-fin) and acetaminophen (uh-see-tuh-MIN-uh-fin), those headache medicines. I had to stay away from them when I was sick.

Yolanda: Wow, I don't think of medicines as toxins.

Kamika: They do have side effects. I heard that if you take too large a dose of just about anything it can reach toxic levels.

Mr. Lee: In many cases, the effects of medicines—good and bad—would last a lot longer if the liver didn't work so hard at breaking them down quickly.

Rick: I'm just glad I'm gonna make it without a liver transplant.

Kamika: It's a good thing, because there's a shortage of organs. I know because they put my cousin on a waiting list just in case. I think kidneys are a little easier to get.

Mr. Lee: Half of the 20,000 or so transplants done in the United States each year are kidney transplants. About one quarter are liver transplants.

Yolanda: Kidney transplants must be more common because a living person can donate a kidney, since you need only one to survive.

Rick: Well, we could never survive with just half a liver!

Yolanda: Yeah, but the liver can do this cool thing. If you take a dead person's liver and put half of it into two people who need livers, the two halves, uh…

Mr. Lee: Regenerate (rih-JEH-nuh-rate). The halves grow back into complete livers. No other complex organ can regenerate. For example, the heart and the brain can't do it.

Rick: That means a living donor can give half a liver, and the half still left will regenerate.

Kamika: That's really amazing! I can't wait to tell my cousin about regeneration, in case she needs a transplant someday.

➢

ANALYSIS

 1. What are some of the functions of the liver?

2. People who have cirrhosis of the liver are usually on a strict diet. They have to be careful of what they eat and drink. Why do you think this is?

 3. How can understanding how your body works help you make decisions about your health?

LABORATORY

You already know the organs in the digestive system. But what exactly do they do? One important function of the digestive system is to break down food into smaller pieces. Only then can the nutrients in the food be absorbed by your body.

When you chew food, **mechanical breakdown** occurs. Most mechanical breakdown occurs in your mouth with help from your teeth and tongue. Some mechanical breakdown continues in your stomach as it churns the food around. During **chemical breakdown**, substances in your digestive system break down food into even smaller particles. Chemical breakdown begins in your mouth, but occurs mostly in your stomach and intestines.

Does it matter if mechanical breakdown occurs? Find out by modeling the process of food breakdown.

CHALLENGE ➡

Why is it important to chew your food?

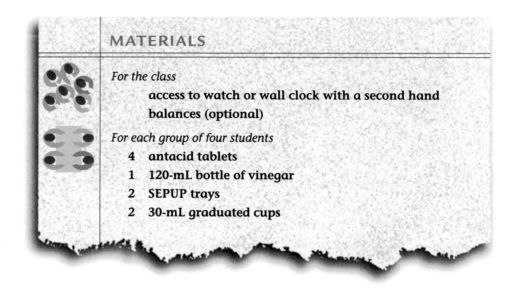

MATERIALS

For the class

    access to watch or wall clock with a second hand

    balances (optional)

For each group of four students

  4 antacid tablets

  1 120-mL bottle of vinegar

  2 SEPUP trays

  2 30-mL graduated cups

 SAFETY

This activity uses vinegar, which has a strong odor.

PROCEDURE

Part One: Testing the Model

The Model	
Material/Process	**Represents**
Antacid tablet	Food
Breaking the tablet	Mechanical breakdown
Adding vinegar	Chemical breakdown

1. Model mechanical breakdown by breaking one antacid tablet into four equal-sized pieces. Imagine that each piece is a small piece of food. Place one piece of food into Cup A of a SEPUP tray.

2. Measure 5 mL of vinegar into a 30-mL cup.

3. Model chemical breakdown by adding the vinegar to Cup A. Observe the reaction until it is over. Then record your observations in your science notebook.

4. Discuss your observations with your group.

Part Two: Designing the Experiment

5. Review the Materials list. With your group, design an experiment to test the effect of mechanical breakdown on the speed of chemical breakdown. In other words, how does the size of your food affect the speed at which chemical breakdown occurs?

When designing your experiment, think about these questions:

- What is the purpose of your experiment?
- What variable are you testing?
- What is your hypothesis?
- What variables will you keep the same?
- What is your control?
- How many trials will you conduct?
- Will you collect qualitative and/or quantitative data? How will these data help you form a conclusion?
- How will you record these data?

6. Record your hypothesis and your planned experimental procedure in your science notebook.

7. Make a data table that has space for all the data you need to record. You will fill it in during your experiment.

8. Obtain your teacher's approval of your experiment.

9. Conduct your experiment and record your results.

10. Create a bar graph of your data. Be sure to label your axes and title your graph.

11. If you have time and additional materials are available, revise your procedure and repeat your experiment.

➢

ANALYSIS

1. **a.** In your experiment, what variables did you keep the same?

   b. Were there any variables (except for the one being tested) that you could not keep the same?

   c. How could you or did you improve the design of your experiment? Explain.

2. **a.** What part of digestion was modeled by breaking the tablet?

   b. What part of digestion was modeled by adding vinegar?

3. How does the size of your food affect the speed at which chemical breakdown occurs? Explain how your conclusions are based on the data collected during your experiment, and whether your hypothesis was supported or disproved.

4. Were your conclusions based on qualitative or quantitative data? Explain.

5. Besides preventing choking, why is it important to chew your food?

READING

Your digestive system is responsible for both mechanical and chemical breakdown. Everything you eat and drink, including medicines, enters your body through this system. What happens as these substances travel through your body?

CHALLENGE

How does your digestive system work?

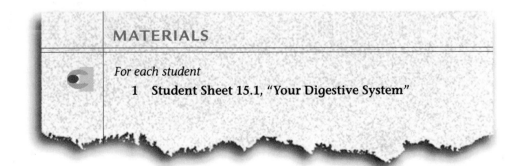

MATERIALS

For each student
1 **Student Sheet 15.1, "Your Digestive System"**

READING

Food Breakdown

Take a moment to look at the diagram of the digestive system in Figure 1. You can think of your digestive system as a long tube that goes through your body. Food is absorbed along this tube. If your body didn't absorb what it needed from the food you eat, everything you swallow would come out the other end! You know that doesn't happen. But do you know why? What are the functions of each part of your digestive system?

The digestive system breaks down food into forms that the body can absorb. This breakdown occurs two ways—mechanically and chemically—and it begins in your mouth. Your teeth begin the process of mechanical breakdown. Chemicals in your saliva begin the process of chemical breakdown.

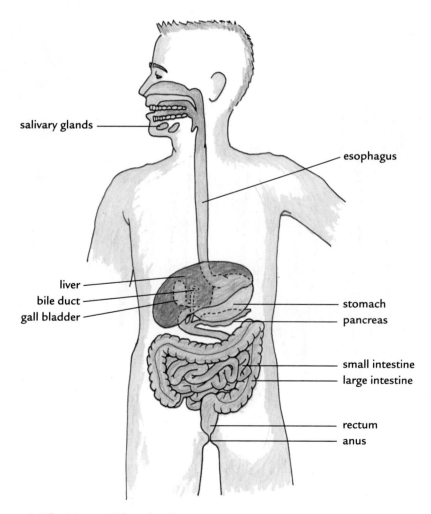

Figure 1: The Human Digestive System

As you swallow, food travels down through your esophagus (ih-SAW-fuh-gus), which is a tube surrounded by muscle. This muscle contracts to help food reach your stomach, a large bag-like organ. Muscles in your stomach wall help to mix the stomach contents. This continues the process of mechanical breakdown. In your stomach, hydrochloric (hi-druh-KLOR-ik) acid—one of the acids used in science laboratories and industries—and other chemicals continue the chemical breakdown of food. The hydrochloric acid in your stomach is so powerful that your stomach is lined with mucus to protect itself. When this lining is absent, ulcers (sores in the lining of the stomach) can form. A high level of hydrochloric acid causes the burning sensation you may feel when you vomit or have indigestion.

STOPPING TO THINK 1

a. How does your mouth contribute to the process of digestion?

b. Explain how your stomach helps break down food.

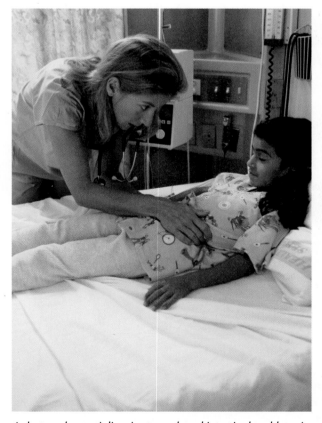

A doctor who specializes in stomach and intestinal problems is called a gastroenterologist (GAS-tro-en-tuh-RAH-luh-jist).

By the time food reaches your small intestine, you wouldn't recognize it anymore! It is a thick pasty mixture. Your small intestine then completes the process of chemical breakdown with help from your pancreas (PAN-kree-us) and liver. As food comes into your small intestine from your stomach, it contains high levels of acid. Your pancreas produces a chemical that reduces this acid level. It also produces chemicals that help break down the proteins and fat found in food. Your liver produces bile, an important mixture that helps break down fat. All of these chemicals combine with the partly broken down food as it travels down your small intestine.

Absorption of Nutrients

Another important process happens in your small intestine, where most of the substances produced by the breakdown of food are absorbed into your blood. After food is completely broken down, we call the pieces **nutrients** (NEW-tree-unts). In the process of **absorption** (ub-SORP-shun), nutrients leave your digestive system and move into your blood, which carries nutrients to the rest of your body. Nutrients are required by all the parts of the body, not just the stomach. The blood acts as the transport vehicle after the stomach has digested food, producing nutrients for all parts of the body.

STOPPING TO THINK 2

a. Explain the relationship between food and nutrients.

b. What role(s) does your small intestine play in digestion?

The fact that most of the final breakdown and nutrient absorption occurs in your small intestine may help explain its length. The average adult small intestine is 5–6 meters (about 15–18 feet) long! This length, plus the folds in the wall of the small intestine (see Figure 2),

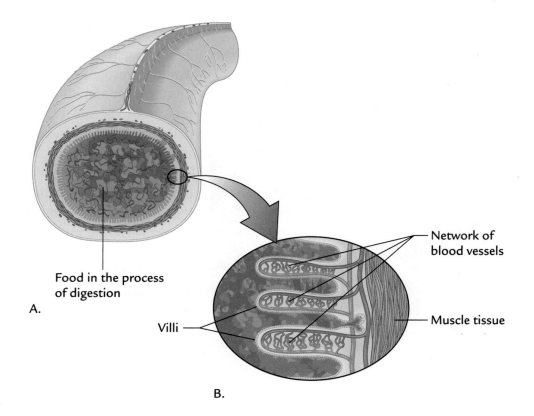

Food in the process of digestion

A.

Villi

Network of blood vessels

Muscle tissue

B.

Figure 2: Cross-Section of Your Small Intestine
Nutrients are absorbed by your blood across the wall of your small intestine. Fingerlike projections from the wall of your small intestine are known as villi (VIL-eye) (singular, villus). Nutrients must pass through villi and the walls of tiny blood vessels to enter your blood.

provides lots of surface area for nutrient absorption. Your blood transports these nutrients to different parts of your body, but first it makes an important stop.

All of the blood that leaves your stomach and intestines goes directly to your liver before traveling throughout the rest of your body. This is because your liver performs two important functions related to digestion, besides producing bile. First, it breaks down toxins such as alcohol and some medicines. (Your blood is later filtered by your kidneys, which excrete liquid wastes and some dissolved toxins as urine.) Second, it processes nutrients into forms that are easier for the rest of your body to use. For example, your liver stores carbohydrates. When you suddenly need energy, it converts these carbohydrates to sugars that your body can use.

STOPPING TO THINK 3

Why does blood travel to your liver before transporting nutrients to other parts of your body?

Getting Rid of Solid Waste

Any material that has not been absorbed by your small intestine continues down into your large intestine, or colon (KOLE-un). In your large intestine, large quantities of water and some remaining vitamins are absorbed into your blood. The remaining unabsorbed material forms a solid waste as it travels through the large intestine, a process that can take 18–24 hours. This solid waste is stored in the rectum (REK-tum) before being pushed out through the anus (AY-nus). What is this solid waste made of? It contains bacteria, substances that your body can't digest, and some remaining water. Bacteria live and grow in your intestines, and they help you in several ways. They break down some plant material that your body can't break down on its own, they make vitamin K, and they help prevent harmful bacteria from finding a home. The trade-off for providing a home for these helpful bacteria is the gas and odors they produce.

....

STOPPING TO THINK 4

The reading describes three components of human solid waste. Which two of these do you think are the main components?

....

ANALYSIS

1. What are some of the functions of the digestive system?

2. Copy the table below. Then fill in the table by placing an "X" to indicate the function(s) of each organ. The first row has been done for you.

Functions of Digestive Organs

Organ (or Structure)	Mechanical Breakdown	Chemical Breakdown	Nutrient Absorption	Water Absorption and Solid Waste Production
Mouth	X	X		
Stomach				
Small intestine				
Pancreas				
Liver				
Large intestine				

3. Imagine taking a bite of a burrito. Follow the beans in the burrito through the process of digestion. Explain what types of changes take place and where each change happens.

4. Most substances are absorbed in the small intestine and not in the stomach. Aspirin is a common exception; it is absorbed in the stomach. Some alcohol is absorbed in the stomach, but most is absorbed in the intestine.

   a. Why would you want medicines, like aspirin, to be absorbed in the stomach instead of the small intestine?

   b. What is the effect of some alcohol being absorbed in the stomach?

5. Take a closer look at the villi of the small intestine (shown in Figure 2b). How do the villi help nutrients move into the blood quickly? **Hint:** What would happen if there were no villi, only a smooth surface?

How does your body use the nutrients it absorbs? Even if you don't exercise, you use (or "burn") energy all the time. Just sitting in a chair for 30 minutes can consume up to 30 Calories! That is why the Food and Drug Administration (FDA) recommends that most kids eat about 2,200 Calories each day. You get these Calories from the food you eat.

Food is made up mostly of carbohydrates, fats, and proteins. One important role for carbohydrates in the body is to provide quick energy. That's why distance runners and other endurance athletes often eat carbohydrates just before a race ("carbo-loading"). Fats, on the other hand, store energy over long periods. Proteins may be used for building muscle and other tissues. They may also be burned for energy when carbohydrates and fats are not available.

How many Calories do *you* need for energy? Consider the energy equation in Figure 1.

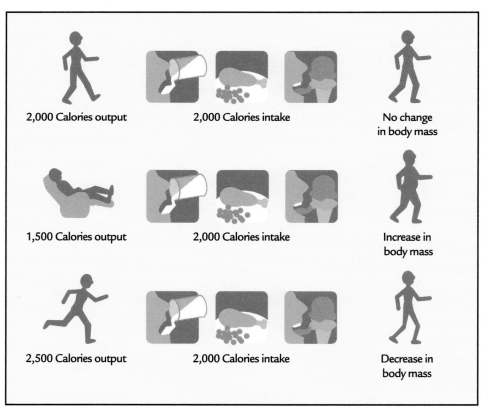

Figure 1: The Energy Equation

CHALLENGE

Can you design and make an energy bar that

• provides between 150–250 Calories?

• provides *no more* than 75 Calories from fat?

• provides *no less* than 25 Calories from protein?

• costs less than $0.50 to make?

• tastes good?

• resists crushing under the weight of a textbook?

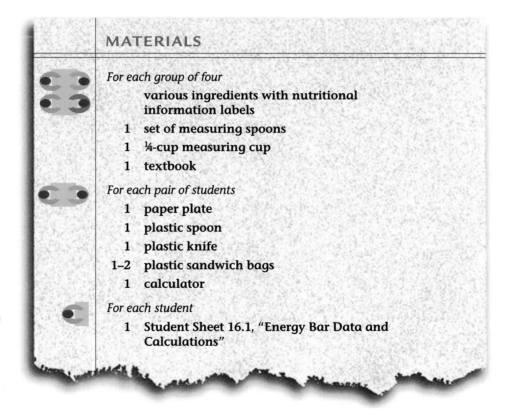

MATERIALS

For each group of four
    various ingredients with nutritional information labels
1   set of measuring spoons
1   ¼-cup measuring cup
1   textbook

For each pair of students
1   paper plate
1   plastic spoon
1   plastic knife
1–2 plastic sandwich bags
1   calculator

For each student
1   Student Sheet 16.1, "Energy Bar Data and Calculations"

SAFETY

Never taste materials or eat or drink in science class unless specifically told to do so by your teacher. Be sure that your work area is clean and free of any materials not needed for this activity. If you are allergic to peanuts or dairy products, or if you have any other health issue, such as diabetes, that limits what you can eat, tell your teacher and do not taste the energy bars in this activity.

Wash your hands before and after preparing the energy bars, and use only food-grade plates and utensils. Do not contaminate foods by licking your fingers or the utensils.

PROCEDURE

Part 1: Designing the Energy Bar

1. Read the energy bar requirements described in the Challenge. Then look at the information provided in Table 1, "Data on Ingredients," on Student Sheet 16.1, "Energy Bar Data and Calculations."

2. Discuss with your group what ingredients you could use to make an energy bar that would meet all of the requirements. Be sure to consider ingredients that are not listed on Student Sheet 16.1.

3. Decide who will bring in the additional ingredients, and gather your materials.

4. Record any additional ingredients you are using in Table 1 of Student Sheet 16.1. Use the nutritional label on the package to fill in the information for each ingredient in Table 1.

5. With your partner, write a recipe for making an energy bar. Include how much of each ingredient you plan to use. **Hint:** Try to use even amounts, such as ¼ cup or 2 tablespoons, to make your calculations easier.

6. Record the ingredients you are using and the amounts of each one in Table 2, "Calculations for My Energy Bar," of Student Sheet 16.1. Be sure to include the units of the amounts you plan to use.

7. Use Table 2 to calculate the number of servings, calories, and cost for each ingredient.

8. Use Table 3, "My Energy Bar Totals," to find out if your energy bar would meet the calorie and cost requirements described in the Challenge.

Part 2: Testing the Energy Bar

9. Follow your recipe and make your first energy bar. Place the bar in a plastic bag and label it with your name and "Energy Bar 1."

10. As directed by your teacher, test the bar for resistance to crushing. Record the results in Table 4, "Evaluating My Energy Bar," on Student Sheet 16.1.

11. Break the bar in half. Leave one half in the bag with the label. Taste the other half and rate it using the scale shown on the left. Record the results of your taste test on Student Sheet 16.1.

12. Use the rest of your data on Energy Bar 1 to complete Table 4 on Student Sheet 16.1. Then discuss with your group how you could improve your energy bar recipe.

13. If you have the time and materials, revise your energy bar recipe to make and test a second energy bar.

Taste Scale		
😄	Good	4
🙂	OK	3
😕	Edible	2
😞	Terrible	1

ANALYSIS

1. What are the similarities between designing a product, like the energy bar, and conducting a scientific investigation? What are the differences?

2. Which requirements did your energy bar meet? Were there any requirements that you could not meet? Explain.

➤

3. The United States Department of Agriculture (USDA) recommends that a person eating a healthy diet consume less than 30% of his or her daily Calories from fat. For a person eating 2,200 Calories every day, this means no more than 660 Calories from fat.

a. Copy and complete a table like the one shown below.

Energy Bar Calories and Fat

	Calories in My Energy Bar	Calories USDA Recommends for One Day	My Bar's Percentage of USDA Recommendation
Total Calories			
Calories from fat			

b. How many of your energy bars could a person eat in one day without exceeding 2,200 Calories, assuming he or she ate nothing else?

c. How many of your energy bars could a person eat in one day without exceeding 660 Calories from fat, assuming that he or she ate nothing else?

4. Explain whether you would recommend your energy bar as a healthy snack. Support your answer with *quantitative* evidence and identify the trade-offs of your decision.

Hint: To write a complete answer, first state your opinion. Provide two or more pieces of evidence that support your opinion. Then consider all sides of the issue and identify the trade-offs of your decision.

5. Look again at Figure 1, "The Energy Equation." Write an explanation of how this figure helps to explain changes in body weight in terms of a balance between food (energy input) and exercise (energy use).

EXTENSIONS

Table 1: Burning Calories	
Activity	**Calories**
Basketball	7
Biking (12–14 mph)	7
Dancing (fast)	5
Football	7–8
Gymnastics	4
Rollerblading	6
Running (6 mph)	9
Sitting (in class)	2
Skateboarding	4
Sleeping	1
Soccer	6
Softball	4
Standing	1
Swimming laps	9
Tennis	6
Walking (4 mph)	4
Watching T.V.	1

Approximate number of calories consumed during 1 minute of activity by a 110-pound person.

1. Imagine that you are in charge of marketing your energy bar. Design an ad to sell it. What further testing and changes would you suggest before your company starts manufacturing your energy bar?

2. *Is your Calorie intake balanced by your level of activity?* Choose a food label from one of your favorite snack foods. How many Calories are provided by one serving of this food? How many servings of this food do you usually eat at one time? Multiply your answers to figure out how many Calories you usually consume when you eat this food.

Look at Table 1, "Burning Calories." Create a weekly activity table to record which activities you do and the number of minutes you do each of them each week. Use Table 1 (or go to the SALI page of the SEPUP website) to calculate approximately how many Calories you burn in an average week.

Choose one of your favorite activities from your weekly activity table. How long would you have to do this activity to use up the number of Calories you usually consume when you eat your favorite snack food (calculated above)?

LABORATORY

Your blood transports the nutrients that you eat to different parts of your body. It also carries oxygen from your lungs to other organs and tissues. With every breath you take, you inhale oxygen and exhale carbon dioxide. Your body uses the oxygen to get energy from food. When your body breaks down food, it produces wastes. One of the wastes is carbon dioxide.

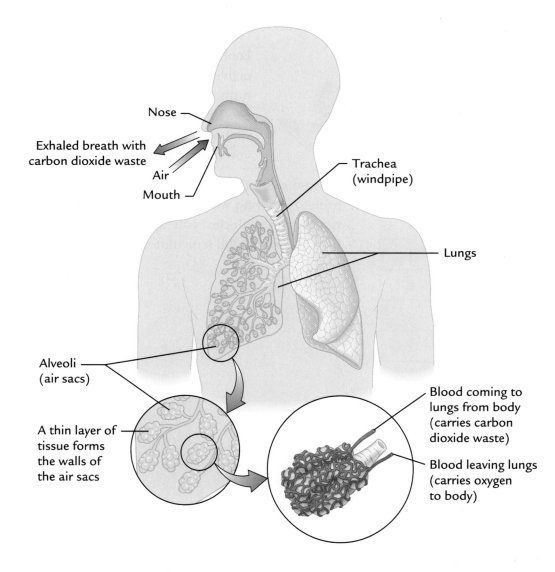

Figure 1: Human Respiratory System

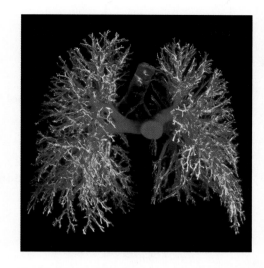

This is a plastic mold of the air ways and air sacs in the lungs. Compare it to the diagram of the lungs on the opposite page.

Indicators (IN-duh-kay-ters) are chemicals that change their appearance in different types of solutions. You will work with the indicator bromthymol (brome-THY-mall) blue, also known as BTB. BTB can be either blue or yellow. When added to a solution containing carbon dioxide, BTB is yellow.

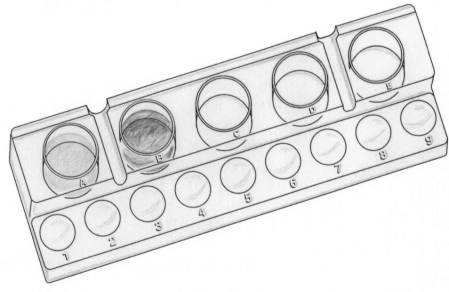

Both solutions contain the indicator BTB. Which cup has a solution containing carbon dioxide?

CHALLENGE

How much carbon dioxide is in your exhaled breath?

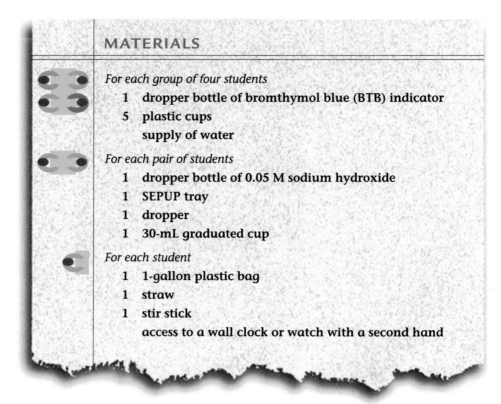

MATERIALS

For each group of four students

1 dropper bottle of bromthymol blue (BTB) indicator

5 plastic cups

supply of water

For each pair of students

1 dropper bottle of 0.05 M sodium hydroxide

1 SEPUP tray

1 dropper

1 30-mL graduated cup

For each student

1 1-gallon plastic bag

1 straw

1 stir stick

access to a wall clock or watch with a second hand

SAFETY

In this activity, you will be blowing through a straw into chemicals. Do not inhale through the straw! Breathe in through your nose and exhale through your mouth. If you accidentally swallow liquid, rinse your mouth thoroughly and drink plenty of water. Be sure to tell your teacher.

PROCEDURE

Part One: Using BTB to Test for Carbon Dioxide

1. Work with your partner to add 5 mL water to each of the five large cups (A–E) of your SEPUP tray. Use the 30-mL graduated cup to measure the water.

2. Add 2 drops of BTB to each cup and stir.

3. Create a data table to record the initial and final colors of the solutions in each cup. Record the initial colors now. Cup A will provide a control.

4. Use your dropper to bubble air into Cup B. Place the dropper into the solution and press the air out of the bulb. Before releasing the bulb, remove the tip from the solution. This will prevent uptake of solution into the dropper. (If you accidentally get solution into the dropper, simply squirt it back into Cup B.) Repeat this for 15 seconds.

5. Record the final color of the solution in Cup B in your data table.

6. Add 3 drops of 0.05 M sodium hydroxide to Cup C. Record the final color in your data table.

7. Unwrap your straw and place one end in Cup D. Take a deep breath, and then gently blow through the straw for 15 seconds. (Remember not to inhale through the straw!) Record the final color of the solution in Cup D in your data table.

8. Have your partner blow through a clean straw into Cup E for 15 seconds. (Remember not to inhale through the straw!) Record the final color in your data table.

9. Add 3 drops of sodium hydroxide to Cups D and E. In your science notebook, record any changes that you observe.

10. Work with your partner to complete Analysis Questions 1 and 2.

Part Two: Using BTB to Measure Carbon Dioxide in Exhaled Breath

11. Work with another pair of students to set up a control:

    a. Measure 10 mL of water using the 30-mL graduated cup.

    b. Add 3 drops of BTB to the graduated cup and stir.

    c. Pour the BTB solution into a large plastic cup. This solution will be the control for every member of your group.

12. Have each person in your group set up his or her own bag of BTB solution:

    a. Measure 10 mL of water using the 30-mL graduated cup.

    b. Add 3 drops of BTB to the graduated cup and stir.

    c. Pour the BTB solution into your own 1-gallon plastic bag.

13. Remove the air from your plastic bag by slowly flattening it. Be careful not to spill any of the BTB solution out of the bag. While keeping the air out of the bag, place a straw in the mouth of the bag. Make an air-tight seal by holding the mouth of the bag tightly around the straw.

14. Be sure you are sitting down. Then fill the bag with air from your lungs by blowing through the straw until the bag is fully inflated. When you finish blowing, pull out the straw. As you pull out the straw, squeeze the bag tightly shut so no air escapes.

15. Holding the bag closed, shake the bag vigorously 25 times.

16. Pour the BTB solution from the bag into a clean, empty plastic cup.

17. *How much carbon dioxide is in your exhaled breath?* You can find out by counting how many drops of sodium hydroxide are needed to make your BTB solution the same color as the control:

    a. Add 1 drop of sodium hydroxide to your plastic cup.

    b. Gently stir the solution and wait at least 10 seconds.

    c. Record in your science notebook that you added 1 drop.

    d. Compare the color of your solution to the control. *Is it the same color as the control for at least 30 seconds?*

    If your answer is no, repeat Steps 17a–d. Be sure to keep track of the total number of drops!

    If your answer is yes, go on to Step 18.

18. In your science notebook, record the total number of drops it took to change your solution back to the same color as the control. Then record your total on the class data table.

19. Draw a bar graph with the class results. Remember to title your graph and label the axes!

EXTENSION 1

Do you exhale more carbon dioxide after you hold your breath? Find out by modifying and repeating Part Two of the Procedure.

ANALYSIS

Part One: Using BTB to Test for Carbon Dioxide

1. What was the purpose of the solution in Cup A?

2. **a.** Which of the solutions in Part One contained carbon dioxide? Support your answer with evidence from your experimental results.

   b. What does this tell you about the exhaled breath of human beings?

Part Two: Using BTB to Measure Carbon Dioxide in Exhaled Breath

3. Review the class data table. What was the range of carbon dioxide in exhaled breath (as measured by drops of sodium hydroxide)? **Hint:** You looked at the range of class data in Activity 9, "Data Toss."

4. Look again at Figure 1, "Human Respiratory System." Considering all the oxygen that has to get into your blood and all the carbon dioxide that has to escape from your blood, why do you think the inside of the lung is structured the way it is?

5. **a.** Were the data collected in Part One qualitative or quantitative? Explain.

   b. Were the data collected in Part Two qualitative or quantitative? Explain.

6. **a.** Look carefully at Figure 1, "Human Respiratory System." What are some of the important structures in the respiratory system?

   b. Explain how gases are exchanged within the respiratory system.

EXTENSION 2

How do you think your body gets more oxygen when you exercise? Do you breathe faster (take more breaths per minute)? Or do you absorb more oxygen from the air with each breath? Use what you learned in this activity to develop an experiment to test your hypothesis.

MODELING

How do the nutrients absorbed from the digestive system get to every part of your body? They travel in your blood through your **circulatory** (SIR-kyu-luh-tor-ee) **system**. In Figure 1, you can see that your blood goes to every part of your body. The main function of your blood is to transport, or carry things around. Blood transports oxygen, nutrients, and wastes such as carbon dioxide. Some organs in your body help get nutrients and oxygen into your blood and also remove wastes. In this activity, you will model the path of your blood as it moves through your body and you will learn what happens along the way.

CHALLENGE ➡️

What does blood do as it travels around your body?

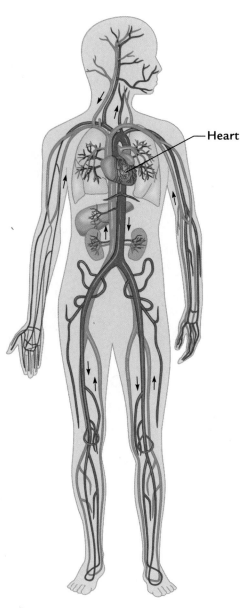

—Heart

Figure 1: Human Circulatory System

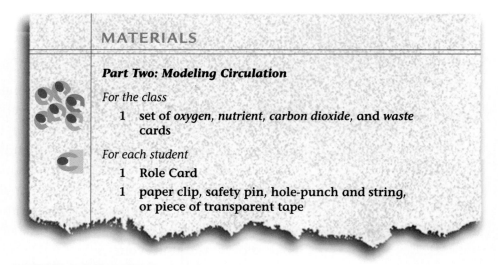

MATERIALS

Part Two: Modeling Circulation

For the class

1   set of *oxygen, nutrient, carbon dioxide,* and *waste* cards

For each student

1   Role Card

1   paper clip, safety pin, hole-punch and string, or piece of transparent tape

PROCEDURE

Part One: Blood Flow

1. Look at Figure 2, "Diagram of Blood Flow." This is a simplified map of how blood travels around your body. Use your finger to trace one of the possible paths of blood flow. Begin on the left side of the heart (on your right) and stop once you reach the left side of the heart again. Be sure to go in the direction of the arrows.

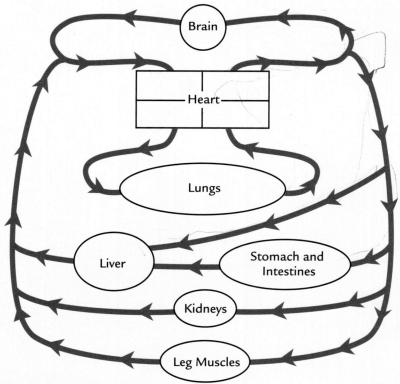

Figure 2: Diagram of Blood Flow

2. In your science notebook, record which organs and structures you passed through in your path.

3. Repeat Steps 1 and 2 by tracing a different path of blood through the human body.

4. Discuss the following questions with your group members:

   • Did everyone trace the same paths? If not, compare the organs (and structures) along the different paths.

   • Which organs does the blood have to pass through each time it goes around the human body?

   • Why do you think blood always has to pass through these organs?

5. Use your discussion and your knowledge of the human body to complete a table like Table 1.

Table 1: Functions of Certain Organs

Function	Organ(s)
Pumps blood	
Brings oxygen into the body	
Carries carbon dioxide out of the body	
Absorbs nutrients	
Removes wastes	

6. As directed by your teacher, share your discussion with the rest of the class.

Part Two: Modeling Circulation

> ## THE MODEL
>
> You will model the way in which your blood transports oxygen, nutrients, and carbon dioxide and other wastes. Each student will receive a Role Card to role play one part of the human body: blood, brain, heart, lungs, stomach and intestines, liver, kidneys, or leg muscles.
>
> In this model, as "blood" flows through the "human body," it will absorb "oxygen" and "nutrients" and carry them to other parts of the body. "Organs" will use the nutrients and oxygen and get rid of carbon dioxide and other wastes by giving them to the blood. Colored cards will represent the four substances (oxygen, nutrients, carbon dioxide, and waste) that are being transported.

7. Your teacher will give you a Role Card. Read your Role Card carefully to see what your job will be. Be sure to note how many *oxygen, nutrient, carbon dioxide,* and *waste* cards you will need to collect before beginning.

8. Collect the cards that you need to begin. Keep *oxygen, nutrient, carbon dioxide,* and *waste* cards in separate stacks so that exchanges can take place quickly. Then attach your Role Card to your clothing.

9. Find your place on the blood flow diagram before you begin the game.

10. As a class, play the Circulation Game.

11. Switch Role Cards before doing more trials.

ANALYSIS

1. Compare this circulation model to the human body. How well did the Circulation Game represent what really happens inside your body?

2. Do all parts of the human body use oxygen and nutrients? Explain your answer.

3. Why does blood flow from the stomach and intestines directly to the liver? **Hint:** Review your notes from Activity 13, "Living with Your Liver," and Activity 15, "Digestion—An Absorbing Tale," for help.

4. What are the functions of the blood as it travels around the human body? Be specific.

5. Look at the diagram below.

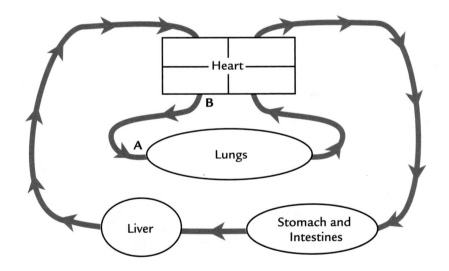

a. Use your finger to trace the path between Point A and Point B, making sure to follow the direction of the arrows. List the organs in the order in which you passed through them.

b. Imagine blood carrying only carbon dioxide and nutrients at Point A. Describe what happens to the blood as it flows from Point A to Point B.

INVESTIGATION

Imagine going on a long hike. How would this exercise affect your body? How could you measure this effect? When you exercise, your heart beats faster to provide more oxygen to the muscles in your body. The harder you exercise, the faster your heart beats. You can measure the speed at which your heart beats by taking your **pulse**, which reflects the contractions of your heart. Each time your heart beats, it sends blood through your body. You can feel this surge when you press on blood vessels near the surface of your skin.

One way you can investigate your level of fitness is by measuring how long it takes for your heart to recover from exercise. The more physically fit a person is, the more efficiently oxygen and nutrients can be transported to the muscles. The faster your heart recovers from exercise, the more fit you are.

CHALLENGE

How can you quantitatively measure your level of fitness?

Did you know that you can have a career studying exercise and helping people get into shape and stay fit? This type of work is known as exercise physiology (fih-zee-AH-luh-jee).

MATERIALS

For the class

    access to a wall clock or watch with a second hand

For each pair of students

  1 **calculator**

For each student

  1 **Student Sheet 19.1, "Pulse Data"**

SAFETY

If you begin to feel dizzy or short of breath during exercise, stop exercising immediately. Do not participate in this activity if you have any condition that prevents you from exercising.

PROCEDURE

Part One: Resting Pulse

You can monitor your pulse at a number of sites. Two convenient sites to use are:

· the artery at the base of the wrist (of either hand)

· the artery at the side of the neck.

1. To take your pulse at rest, sit comfortably. (The best and most accurate time to measure resting pulse is immediately after you wake up in the morning, before you get out of bed. You can check your resting pulse at that time to see how it varies from the values taken in class.)

2. Use the first two fingers of one hand to locate your pulse at the base of your wrist.

3. Measure your pulse for 15 seconds. Have your partner keep track of the time.

4. Record your data for Trial 1 on Student Sheet 19.1, "Pulse Data."

5. Repeat Step 3 two more times and record your data for Trials 2 and 3.

6. Repeat Steps 1–5 for your partner.

7. Calculate your pulse for 60 seconds (1 minute) by multiplying each 15-second pulse by 4.

8. Calculate your average resting pulse per minute. Do this by adding all the numbers in the column titled "60-Second Pulse." Then divide your total by 3. Record your average resting pulse per minute on Student Sheet 19.1.

Part Two: Recovery Time

9. As discussed in class, exercise for 5 minutes. You should begin to feel your heart beating faster. **Hint:** If you can have a normal conversation while exercising, you are not exercising hard enough. If you cannot talk at all, you are exercising too hard.

10. After 5 minutes, stop exercising and sit down. Immediately begin taking your pulse every 30 seconds for the next 5 minutes. Record your 15-second pulse every 30 seconds on Student Sheet 19.1.

11. Calculate your 60-second pulse for each time period by multiplying each 15-second pulse by 4.

ANALYSIS

1. What happened to your breathing rate during exercise? Discuss what was happening inside your body that caused this to happen.

2. What caused the difference between your resting pulse and your pulse after exercise? In other words, what was happening inside your body that caused your pulse to change?

3. a. *Recovery time* is the time it takes for your pulse to return to within 20% of your resting pulse. In order to measure your recovery time, you must first know when you are within 20% of your resting pulse. Calculate this value by multiplying your resting pulse by 1.2.

Resting pulse X 1.2 = _____ beats/min

b. Look at your 60-second pulse values in Section III, "Recovery Time," of Student Sheet 19.1. How many minutes after you stopped exercising did it take you to return to within 20% of your resting pulse? (This is your recovery time.)

4. Use the guide below to create a line graph of pulse vs. time. Plot the data from your 60-second pulse during the recovery period.

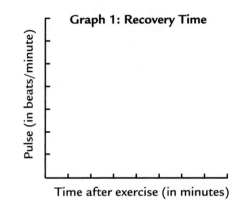

5. If you improved your level of physical fitness, would you expect your resting pulse to increase or decrease? Explain.

6. What do you predict would happen to your recovery time if you exercised at least three times a week for a month?

EXTENSION

What exercises do you enjoy? Identify an exercise that you could practice to improve your recovery time. Or go to the SALI page of the SEPUP website for some ideas. Then develop a plan to exercise regularly for the next month. You should exercise at least three times a week for at least 20 minutes each time. After exercising for a month, repeat Part Two of the Procedure. What effect does regular exercise have on your recovery time?

ROLE PLAY

Reproduced with permission from the American Lung Association.
"Shop Smart with Heart" ©2001, American Heart Association.

Have you ever heard of the American Heart Association, the American Cancer Society, or the American Lung Association? Each of these organizations raises money for research to fight specific diseases. For example, the American Heart Association focuses on diseases of the cardiovascular (CAR-dee-oh-VAS-kyu-lur) system, which you know as the circulatory system. People at the Association must decide how to spend the money they raise to fight heart disease. Sometimes they develop brochures like the ones shown here to help people learn about a disease. These brochures are one way organizations work to improve public health. There are many other ways.

Would you like to develop a public health campaign? Some people do this as a career. Imagine you have to decide how to spend a limited amount of money on public health. What would you do?

CHALLENGE

What are the trade-offs involved in choosing research, treatment, or education as approaches to promote public health?

MATERIALS

For each student
1 **Student Sheet 20.1, "Requests for Funding"**

PROCEDURE

1. Watch the role play of the story of Great-Aunt Lily's will. As you watch, decide what you would do if you were in Julia's place.

2. Use Table 1, "Awarding Money," on Student Sheet 20.1, "Requests for Funding," to record which organizations have requested money and how much each organization has requested.

3. In Table 1, identify whether the organization has requested money for education, research, or treatment.

4. Complete Table 1 by deciding how much money you will award to each organization. Remember, your total cannot exceed 1 million dollars.

5. Calculate the percent of the 1 million dollars you awarded for education, research, or treatment by completing Table 2, "Percent Awarded," on Student Sheet 20.1.

GREAT-AUNT LILY'S WILL

"I just can't decide what to do," Julia moaned as she looked at the papers all over her desk.

Hearing her complain, James, Trina, and Harold poked their heads around the door. "What's up?" asked Harold.

Julia pointed to all the documents on the desk. "In front of me I supposedly have all the information I need to make a decision. You know Great-Aunt Lily left 1 million dollars to 'help fight heart disease.' That's a lot more complicated than it sounds. First, there are reports that say one way to reduce heart disease is to educate people about heart disease and prevention. Second, researchers want money to study various heart problems.

They hope to develop new and better treatments. Finally, some people need help paying for medical procedures."

"Giving money to lots of good causes—sounds like fun to me!" Trina remarked.

"Well, that's true," Julia sighed. "But 1 million dollars isn't enough for such a big problem. Almost 25% of the population has some type of heart disease. In fact, heart disease has been the top killer in the United States every year since 1900! With heart disease so common, a lot of people need money."

James picked up a letter off the desk. "This looks easy. This organization called Making A Difference helps people who can't pay for surgeries. They want $500,000 to help three families pay for the costs of heart surgery…. How sad—one of the families has a little girl who was born with a hole in her heart. That's been fixed, but she needs more surgery. And in this other family, a dad needs a heart transplant."

"I know," replied Julia. "It seems like Making A Difference could really change the lives of these three families. But funding public education on heart disease could reach thousands of people. Providing people with information could mean that they take better care of themselves. This could reduce their risk of heart disease. More people would stay healthier and the cost of medical care would be reduced. The cost of heart diseases and strokes in 1998 alone was 274.2 BILLION dollars!"

"Whoa, that's a lot of money." Harold began to leaf through some of the folders. "But how does education help? No one pays attention to that stuff. After all, everyone believes it'll never happen to them."

"I don't think that's true," said Trina. "Look at all the people who quit smoking when they realized what it does to their heart and lungs. Many people don't even know the early warning signs for heart disease. They don't know what to do when someone has a heart problem, like a heart attack. But people do pick up and read brochures while waiting to see the doctor."

"Exactly." Julia went on. "Producing a million brochures for health service agencies, like doctors' offices, hospitals, and community health clinics, would cost around $400,000. Sending out brochures could potentially help a million people. It might prevent hundreds of cases of heart disease!"

"I see where you got those numbers," James said as he held up a folder. "A group called HeartSmart has asked for $400,000 to do exactly what you just said."

Harold held up another file. "Here's another organization that wants money for public education. Project Heart needs $685,000 to produce and air 15-second spots on television. Their request doesn't say how often or on what channel, but having informational spots on TV is a cool idea! I bet a lot of future heart problems could be prevented that way. In fact, we'd be helping the next generation of adults."

"Harold, you're not helping!" exclaimed Julia. "I haven't even mentioned research. Scientists around the country are trying to identify the causes of heart disease. Other scientists are researching better treatments for heart problems. Without money, some of this research will end. Who knows which research project might result in the next breakthrough!"

"How much does a research project cost?" Harold asked.

Some research scientists work to develop new drugs or new treatments for disease.

"Research is expensive," replied Trina as she looked at a report. "CardioResearch requested $807,000. They want to continue their research on a drug that would help people with high blood pressure."

James read aloud from another letter. "University Research Hospital would like exactly 1 million dollars to continue its work developing better surgical procedures for blocked blood vessels. Hmmm...that would really help a lot of people whose arteries are blocked. You probably know someone who has had bypass surgery to deal with that problem."

Harold had an idea. "How about splitting up the money? Maybe you could give some of these people only part of what they asked for. Then they might be able to get money from other places as well."

"Thanks, Harold," said Julia. "Now you know exactly where I started: What's the best way to spend the 1 million dollars? Who should get the money? And how much?"

ANALYSIS

1. Discuss with your group why you decided to award the money as you did.

2. What additional information (or types of information) would have helped you make a better decision?

3. What are the trade-offs of choosing only one way—education, research, *or* treatment—to deal with a public health problem such as heart disease?

4. **Reflection:** What effect do you think public education, such as short television spots, has on people's behavior? Support your answer with evidence from your own personal experience.

You know that your blood transports oxygen and carbon dioxide from one part of your body to another. Your heart is the pump that keeps your blood moving. Your heart pumps a lot like a mechanical pump. You will find out more about the structure and function of your heart by using a mechanical pump as a model.

CHALLENGE

What type of pump is better for pumping water? What does this tell you about the structure of your heart?

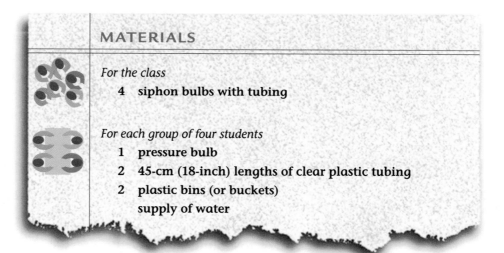

MATERIALS

For the class
   4 **siphon bulbs with tubing**

For each group of four students
   1 **pressure bulb**
   2 **45-cm (18-inch) lengths of clear plastic tubing**
   2 **plastic bins (or buckets)**
      supply of water

SAFETY

Do not attempt to use your mouth to start a pump. During this activity, water may spill on the floor, so walk carefully.

PROCEDURE

1. Fill one of the bins ¾ full of water. (Your teacher may have already done this.)

2. Connect two pieces of plastic tubing to the pressure bulb, one on each end.

➢

Figure 1: The Pump Set-Up

3. Place one end of the tubing in the container of water and the other in the empty container, as shown in Figure 1.

4. Use your pump (pressure bulb) to transfer the water from one container to the other. (You may have to hold the bulb at or below the level of the water to get it started. Or you can squeeze the bulb and place your finger over the open end of the tubing while releasing the bulb to allow water to flow into the bulb.)

5. Borrow a siphon bulb from your teacher and try to create a pump that can transfer water from one container to the other.

6. Record how you made one or both pumps work. You can draw pictures to show what you did.

7. Try to figure out how the pressure bulb works. **Hint:** You can remove the tubing, shake the bulb, and listen to both ends as you squeeze air through the bulb.

8. Watch your teacher demonstrate how a siphon bulb works.

ANALYSIS

1. If you cut lengthwise through each bulb, the outer walls of the bulbs will look like the diagrams below. These are called lengthwise *cross-sections*. A cross-section gives a view inside a sliced object.

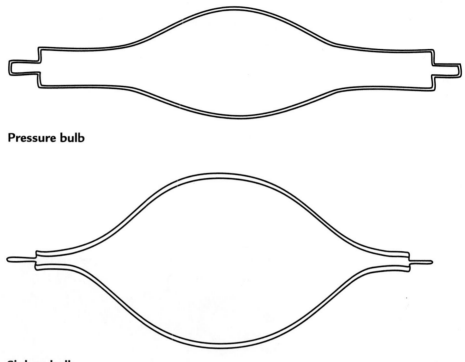

Pressure bulb

Siphon bulb

> **a.** What do you think is inside each bulb? Draw a cross-section of both the siphon bulb and the pressure bulb, as shown above. Complete the cross-section by drawing and labeling what you think is inside each bulb.
>
> **b.** On your drawings of the bulb cross-sections, add arrows showing which way water flows inside the bulb.
>
> **c.** What made one bulb work better than the other?

2. Which type of bulb is better at pumping water? Why?

3. Your heart pumps blood around your body. Would you expect it to work more like the siphon bulb or the pressure bulb? Explain.

LABORATORY

Did you know that your heart is a **muscle?** It's a muscle that works constantly—every minute of every day of every month of every year—for your whole life. You never feel it getting tired, but how hard is it working? Did you know that your heart may beat up to 3 billion times during your lifetime? In this activity you will explore how hard your heart works.

CHALLENGE

How can you measure how hard your heart muscle works?

MATERIALS

For each group of four students

1   **pressure bulb**
2   **45-cm (18-inch) lengths of clear plastic tubing**
2   **plastic bins (or buckets)**
1   **1-liter bottle or other measuring container**
     supply of water
1   **meter stick (optional)**
     paper towels
     access to a wall clock or watch with a second hand

SAFETY

During this activity, water may spill on the floor, so walk carefully.

PROCEDURE

1. Measure and record your resting pulse (in beats per minute). **Hint:** For help on how to do this, review Activity 19, "Heart-ily Fit."

2. Fill one of the bins or buckets ¾ full of water. (Your teacher may have already done this.) The other bin should remain empty.

3. Connect two pieces of plastic tubing to the pressure bulb, one on each end. Place one end of the tubing in each bin.

4. Have members of your group help you keep time and hold the tubing in place as you do the following:

   a. Practice using one hand to pump water from one bin to the other. When you start, be sure to hold the bulb at the same level as the water.

   b. Pour all the water back into one bin so that one bin is again ¾ full of water and the other bin is again empty.

➢

Figure 1: Heart Output

Height (feet and inches)	Volume of Blood (liters per minutes)
4'0"	3.00
4'1"	3.13
4'2"	3.25
4'3"	3.38
4'4"	3.50
4'5"	3.63
4'6"	3.70
4'7"	3.75
4'8"	3.88
4'9"	4.00
4'10"	4.13
4'11"	4.20
5'0"	4.25
5'1"	4.38
5'2"	4.50
5'3"	4.63
5'4"	4.75
5'5"	4.88
5'6"	4.95
5'7"	5.00
5'8"	5.13
5'9"	5.25
5'10"	5.38
5'11"	5.45
6'0"	5.50

c. Use one hand to try to pump the bulb the same number of times as your resting pulse in 1 minute. For example, if your resting pulse is 76 beats per minute, try to pump your bulb 76 times in 1 minute.

d. Record the number of times you were able to pump the bulb in one minute.

e. Use the measuring container to measure the amount of water in the second (initially empty) bin. This is the **volume** you pumped. Record this value in your science notebook.

f. Record in your science notebook how your hand felt after you finished pumping the water.

5. Have each member of your group repeat Step 4 as you help them keep time and hold the tubing in place.

ANALYSIS

1. Compare the pressure bulb model to what you know about your heart. In what ways do you think the pressure bulb is a good model for your heart? What are the weaknesses of the pressure bulb as a model for the heart?

2. Use Figure 1 to find out how much blood your heart pumps per minute based on your height.

   a. Record the volume of blood (in liters) pumped by your heart each minute.

   b. Compare the amount of blood your heart pumps each minute to the amount of water you were able to pump: Was it more? Was it less? By how much?

3. Describe how hard your heart works by using quantitative and qualitative data from this activity. **Hint:** Be sure to look at your notes from this activity.

4. Why do you think that exercising regularly decreases your heart rate?

READING

In Activity 21, "Inside A Pump," you observed two bulbs: one contained valves and one did not. The valves prevented the backward flow of water, which helped pump water from one container to the other. The human heart has **valves** that help blood flow in one direction. Where are these valves located? Why are they necessary?

CHALLENGE

How does your heart work as a double pump?

MATERIALS

For each student
1   Student Sheet 23.1, "Heart Diagram"
     red and blue colored pencils

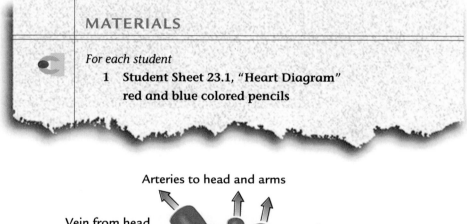

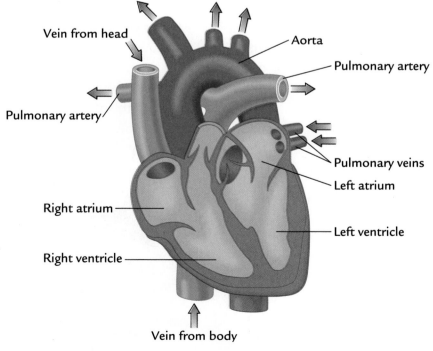

Arteries to head and arms

Vein from head

Aorta

Pulmonary artery

Pulmonary artery

Pulmonary veins

Left atrium

Right atrium

Right ventricle

Left ventricle

Vein from body

Figure 1: Heart Diagram

READING

Your heart is made up of four chambers that work as two pumps. The right side of your heart acts as a pump that pumps blood to your lungs. The left side of your heart acts as another pump that pumps blood to all other parts of your body. Look at Figure 1 on the previous page. Blood always enters your heart through either of the two chambers known as atria (AY-tree-uh)—the plural for **atrium** (AY-tree-um). Blood is pumped out of your heart through the ventricles—plural for **ventricle** (VEN-trih-kul).

STOPPING TO THINK 1

On Student Sheet 23.1, "Heart Diagram," use your finger to trace the flow of blood through the right side of the heart. Notice where the blood is coming from and where it is going. Repeat this process for the left side of the heart.

There are valves located between your atria and ventricles. There are also valves between your ventricles and the large vessels that lead to the lungs and the rest of your body. The valves open to let blood flow from one structure to another. The valves close to prevent the blood from flowing backward.

When you listen to your heartbeat, you can hear a *lub-dub* sound. The first part of that sound (*lub*) is the sound of the valves between the atria and ventricles closing. The second part of that sound (*dub*) is the sound of a second pair of valves closing. One of these valves is located between the left ventricle and the blood vessel carrying oxygenated blood to the body's organs. The other valve is located between the right ventricle and the blood vessel carrying deoxygenated blood to the lungs.

Medical and scientific illustrators enjoy careers that blend science and art.

STOPPING TO THINK 2

Look carefully at the heart on Student Sheet 23.1. Identify the location of the heart valves. Circle each valve. Then label the valves that produce the *lub* sound and the valves that produce the *dub* sound.

Blood travels through your body in tubes of various sizes. These tubes are known as **blood vessels** (VEH-suls). Blood vessels form a network of tubes throughout your body. At various points in the network, blood vessels are called arteries, veins, or capillaries. Remember finding your pulse in Activity 19, "Heart-ily Fit"? You can feel your pulse as your heart pumps blood through your arteries. **Arteries** (AR-tuh-rees) carry blood away from your heart. Most arteries carry oxygen-rich blood to the organs. The largest artery in your body is the aorta (ay-OR-tuh). **Veins** (VANES) carry blood back to your heart. Most veins carry blood with lower levels of oxygen and higher levels of carbon dioxide (picked up from the organs). Even though the blood in your veins is a deep red, veins look dark purple or blue under the skin of your wrists and under your tongue.

STOPPING TO THINK 3

a. Color your heart diagram on Student Sheet 23.1. Use red for areas that contain blood carrying higher levels of oxygen (and lower levels of carbon dioxide). Use blue for areas that contain blood carrying higher levels of carbon dioxide (and lower levels of oxygen).

b. Recall that arteries carry blood away from your heart and that most arteries carry blood with higher levels of oxygen. Look carefully at the pulmonary (PULL-muh-nair-ee) arteries on Student Sheet 23.1. Explain how these arteries are different from most arteries in your body. **Hint:** Think about the blood they are transporting.

c. Recall that veins carry blood to your heart and that most veins carry blood with lower levels of oxygen. Look carefully at the pulmonary veins on Student Sheet 23.1. Explain how these veins are different from most veins in your body. **Hint:** Think about the blood they are transporting.

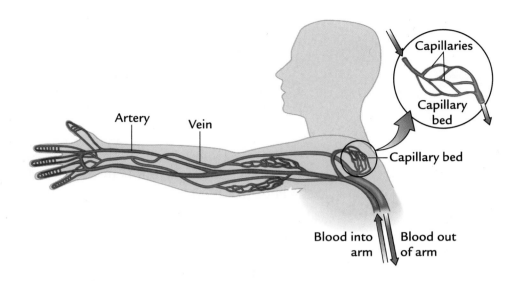

Figure 2: Diagram of Circulatory System (in arm)

Blood leaving the left side of your heart flows into arteries that carry oxygen to all parts of your body. Arteries become smaller and smaller until they become capillaries. **Capillaries** (KA-puh-lair-ees) are blood vessels with walls so thin that oxygen, nutrients, and wastes can pass back and forth. Each capillary is very small and there are many of them. They provide lots of surface area for oxygen to move to the tissues and carbon dioxide wastes to move out of the tissues. As the capillaries widen again, they become veins.

ANALYSIS

1. How is the structure of the heart related to its function?

2. What structures prevent blood in the ventricles from backing up into the atria? Why is it important for your heart to have these structures?

3. Explain what is meant by the statement: "The heart is two pumps." You may want to draw a diagram to support your explanation.

EXTENSION

How did scientists learn about how the human body works? During the 1620s, one scientist, William Harvey, performed dissections to investigate the circulatory system. Research his work to find out more.

LABORATORY

In Activity 18, "The Circulation Game," you modeled the flow of blood through the heart and around the body. In Activity 21, "Inside A Pump," you saw that the heart resembles a pump. Look at the photo below. How well does this model represent the structure of the heart? How well does this model represent the function of the heart?

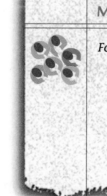

CHALLENGE ➡

How can you construct a model to show how the heart pumps blood to the lungs and the rest of the body?

MATERIALS

For the class

pressure bulbs

45-cm (18-inch) lengths of clear plastic tubing

plastic bins (or buckets)

supply of water

red and blue colored pencils

siphon bulbs with tubing

PROCEDURE

1. Review the Materials list on the previous page.

2. With your group, design a model using the materials to show how the heart pumps blood to the lungs and the rest of the body. **Hint:** Think about how you modeled the flow of blood through the body in Activity 18, "The Circulation Game."

   • Work together to agree on a design.

   • Listen to and consider the explanations and ideas of other members of your group.

   • If you disagree about a factor with others in your group, explain to the rest of the group why you disagree.

3. Draw a diagram of your model. Label what each part of your model represents.

4. Share your model with another group.

5. Work together to choose one model to test.

6. Set up and test the model.

7. Use your diagram to explain your model to the class.

ANALYSIS

1. How well did your group's original model work? What changes did you make to improve it? Discuss how your design showed how the heart pumps blood to the lungs and the rest of the body.

2. **a.** Draw a diagram of your final model. Use arrows to show which way the water flowed.

   b. Label the parts of your model that represented various organs, structures, or systems of the human body.

   c. Use a *red* colored pencil to identify which tubing contained blood carrying more oxygen and less carbon dioxide. Use a *blue* colored pencil to identify which tubing contained blood carrying less oxygen and more carbon dioxide.

3. Compare the different models that were presented. Which design(s) modeled the function of the heart the best? Explain.

4. **Reflection:** How does modeling help you understand how things work?

READING

Your heart beats every minute of your life. If your heart stops beating, you only have about four minutes before severe brain damage and death occur. This makes it difficult for doctors to treat heart problems. Imagine trying to perform surgery on beating tissue when any cut forces blood out of the heart. How have doctors overcome this challenge? In this activity, you will learn about advances in heart surgery that have saved many lives.

CHALLENGE

What are some of the risks in developing new treatments for heart problems?

MATERIALS

For each student
  1 **Student Sheet 25.1. Heart Surgery Timeline**

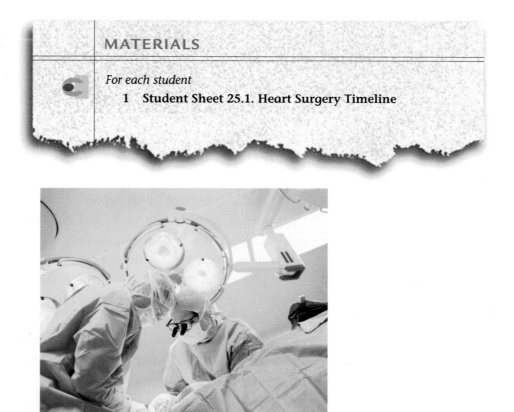

READING

Dr. Daniel Hale Williams performed the first open-heart surgery in 1893, before x-rays, blood transfusions, and antibiotics. The patient lived for over 20 years after the operation.

Early Heart Surgery

In 1893, a man arrived at a Chicago hospital with a knife wound in his chest. The chief doctor saw that the man was bleeding to death, and he made a risky decision: open the chest of the man, and repair the knife wound. In 1893, any operation on an internal organ ran a high risk of infection and death because sanitary procedures were poor, even in hospitals. Luckily for this patient, the doctor, Daniel Hale Williams, insisted on sanitary procedures. Probing inside the patient's chest, Dr. Williams found that the sac surrounding the heart had been cut. He sewed it up and became one of the first doctors to successfully operate on the heart. The patient was still alive 20 years later! Dr. Williams's sanitary procedures also set a new standard for surgeries.

Nearly 50 years later, most doctors considered surgery on the heart itself too dangerous to perform. But then, during the Second World War, a U.S. Army surgeon, Dr. Dwight Harken, created a technique to remove shrapnel from soldiers' hearts. He could make a small incision in the side of the heart and reach in with his fingers or a clamp to remove the shrapnel. When the war ended, doctors used similar methods to repair damaged heart valves. The discovery of antibiotics also improved patients' chances of surviving surgery.

Open Heart Surgery

To do more extensive heart repairs, doctors had to figure out how to stop the heart during surgery. The constant movement as the heart beat made it nearly impossible to operate on. Simply stopping the heart, however, was not the answer. The four minutes a patient can live without oxygen was not enough time to repair heart problems. Two things happened that allowed heart surgery to progress.

First, Dr. Wilfred Bigelow, a surgeon at the University of Minnesota, suggested cooling patients during heart surgery. He thought this might reduce the patients' need for oxygen. In 1952, surgeons at the University of Minnesota lowered a patient's body temperature to 81 degrees Fahrenheit (27 degrees Celsius). The surgeons were able to extend operating time to 10 minutes! That helped in some cases, but it was not enough time for more complex surgeries.

A heart-lung machine

The second breakthrough came in 1958, when a heart-lung machine was perfected. This machine delivered oxygen-rich blood to a patient's body while the heart was stopped. Doctors could perform longer operations on the stopped heart without risking brain damage or death. People with serious heart defects could now have their hearts repaired.

But what about people with heart disease whose hearts could not be repaired? If they were going to survive, they would need a heart transplant.

Artificial or Human Heart?

In the 1950s in Houston, Texas, Dr. Michael DeBakey and Dr. Denton Cooley developed new methods for operating on the heart. After working together for several years, however, they decided they could no longer get along with each other. In the early 1960s they went separate ways, although they both still worked in the same city.

During the early to mid-'60s Dr. DeBakey worked with heart failure patients. These patients' hearts were no longer strong enough to pump blood. Dr. DeBakey knew the left ventricle of the heart did the actual pumping of the blood to the body. He thought that he could help these patients if he could make their left ventricles pump blood. And so he invented a small pump that he could put in a person's left ventricle to help the heart pump blood. The device, with improvements, is still used today. His bigger dream, however, was to develop an artificial heart that could replace a diseased heart.

Dr. Michael DeBakey holding artificial heart grafts

On the other side of the world in 1967, a South African surgeon, Dr. Christiaan Barnard, performed the first human heart transplant. He removed a diseased heart from a patient and replaced it with a heart from a young woman who had died in a car accident. The patient lived 18 days. Shortly after that, a heart transplant was performed in the United States, but that patient lived only six hours. However, Dr. Barnard's next heart-transplant patient lived for 18 months. While the survival time of these patients seems short today, it was a major advance at the time. Transplant surgeons were only allowed to take patients who had no hope for survival. In many cases these people were days away from death. They had to consent to the risky, experimental surgery. Today, more than 2,000 successful heart transplants are performed each year.

About this time, Dr. DeBakey received a grant to develop an artificial heart. He and his team built an experimental model in his laboratory, but tests on animals were not successful. He knew it would be a long time before he could fulfill his dream and use the artificial heart to help a human patient.

In 1969, his former partner, Dr. Cooley, transplanted an artificial heart into a man. The man survived for three days until a donor heart became available. Dr. Cooley claimed he and another doctor from Dr. DeBakey's team had invented the artificial heart. Dr. DeBakey claimed that Dr. Cooley took the heart from his lab. Although an investigation was conducted, no one knows what really happened. This dispute led to a 40-year feud between Dr. DeBakey and Dr. Cooley.

The first permanent artificial heart, designed by Dr. Robert Jarvik in Utah, was implanted in a man in 1982. The Jarvik heart was connected to a control unit the size of a shopping cart. Although it was considered a "permanent" heart, a person receiving it could not do much of anything. Other downsides were that many people with artificial hearts developed blood clots, strokes, or other serious problems.

The Problem of Rejection

The body's immune system fights bacteria and other foreign objects. This keeps people healthy. However, if the immune system recognizes a transplanted organ as a foreign object, it attacks the organ. Drugs were developed to decrease the immune response, so a person's body was less likely to reject the new organ. Unfortunately, these drugs also lowered patients' ability to fight off bacteria and other organisms. In 1983, a new drug was approved for transplant patients. It was more effective than previous drugs and had fewer side effects. This increased the survival rate of organ transplant patients. Today, doctors carefully match blood type and tissue type between organ donor and organ recipient. This further improves a patient's chances of living for a long time.

Transplants Today

The major problem for people who need a heart transplant today is that there are not enough donors. In the United States, about 4,000 people need a heart transplant each year. Yet only about 2,200 hearts become available. About 15% of people who are waiting for a transplant die before they can get a donor heart. Artificial hearts can keep people alive for a short time until a donor heart is found. However, the risks of infection or blood clots from artificial hearts are still high.

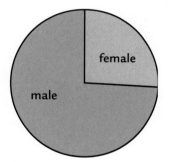

Transplant Recipients by Age	
Age	Percent
Under 18	14%
18–34	11
35–49	20
50–64	44
Over 64	11

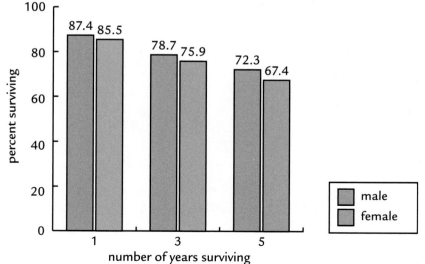

ANALYSIS

1. What is the age range of most transplant patients?

2. What is a heart transplant patient's chance of survival after:

   - one year?

   - three years?

   - five years?

3. Compare the percentages of male and female transplant patients. Why do you think there is a difference? Explain.

4. Why did the early heart transplant patients agree to a transplant when it was so risky?

5. What are the challenges that had to be overcome to develop new surgeries for heart problems?

6. **Reflection:** A person can sign up to be an organ donor when he or she receives a driver's license. Would you be willing to sign up to be an organ donor? Explain.

EXTENSION

Go to the *Science and Life Issues* page of the SEPUP website to link to sites with more information on the history of heart surgery.

INVESTIGATION

How do you know if your heart is healthy? The first thing a doctor usually does is listen to your heart through a stethoscope (STEH-thuh-skope). As you learned in Activity 23, "Heart Parts," your heartbeat sounds like a double beat that is commonly described as *lub-dub*. When blood does not flow normally through the heart, abnormal heart sounds can result.

CHALLENGE

How do heart sounds tell you what is happening inside your circulatory system?

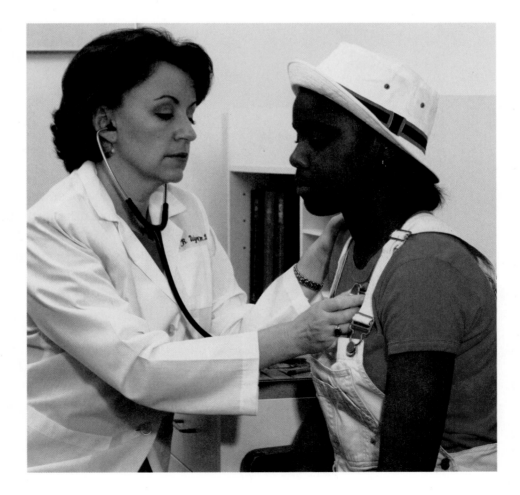

PROCEDURE

Part One: Heart Problems

1. As directed by your teacher, find out more about how the heart works.

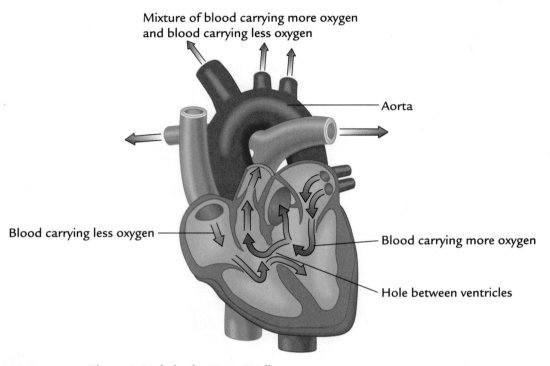

Mixture of blood carrying more oxygen and blood carrying less oxygen

Aorta

Blood carrying less oxygen

Blood carrying more oxygen

Hole between ventricles

Figure 1: Hole in the Heart Wall

Each person is born with a hole between the left and right side of the heart. This hole usually closes just after birth. But sometimes the hole doesn't close properly. In the past, children born with a hole in the heart would eventually die. Now, with the help of modern surgery, these heart problems can often be repaired.

STOPPING TO THINK 1

Look carefully at Figure 1, "Hole in the Heart Wall." What happens to blood with higher as compared to lower levels of oxygen when there is a hole between the ventricles of the heart?

Think about the function of the valves in the heart pump model. What would have happened if the valves in the pressure bulb didn't work? Some of the water would have flowed backward into the bulb. This would have reduced the forward flow of the water. If heart valves don't work properly, the heart cannot pump blood efficiently and the body does not receive enough oxygen. This can cause breathlessness, exhaustion, and in severe cases, death. Heart valve problems can occur at birth or can be the result of aging or disease.

Compare the normal *lub-dub* heart sound to Figure 2, "Contractions of the Heart." The first sound (*lub*) occurs when the valves between your atria and ventricles close. As your ventricles contract, these valves close to prevent your blood from flowing back into your atria. The second sound (*dub*) occurs when the valves leading to your arteries close and prevent your blood from flowing back into your ventricles.

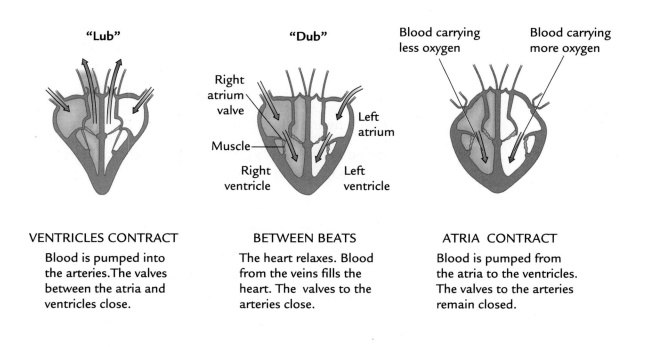

"Lub"

VENTRICLES CONTRACT

Blood is pumped into the arteries. The valves between the atria and ventricles close.

"Dub"

Right atrium valve

Muscle

Right ventricle

Left atrium

Left ventricle

BETWEEN BEATS

The heart relaxes. Blood from the veins fills the heart. The valves to the arteries close.

Blood carrying less oxygen

Blood carrying more oxygen

ATRIA CONTRACT

Blood is pumped from the atria to the ventricles. The valves to the arteries remain closed.

Figure 2: Contractions of the Heart

STOPPING TO THINK 2

Look carefully at Figure 2, "Contractions of the Heart." Discuss the following questions with your group:

- When the left ventricle contracts, where should the blood flow?

- When the left ventricle contracts, the valves between the left atrium and ventricle close. Why? What will happen to blood flow if these valves don't close properly?

- Some heart diseases cause the heart valve openings to narrow. What will happen to blood flow if the valve between the left atrium and ventricle cannot open all the way?

- When the left ventricle relaxes and fills with blood, the valve leading to the aorta closes. What will happen to the blood flow if this valve doesn't close properly?

Part Two: Diagnosis

2. Listen to a series of normal and abnormal heart sounds. As you listen, think about the following questions:

- Do you hear extra beats in the abnormal heartbeat?

- Do you hear a whooshing or echo sound?

- Do you hear an unusual sound during contraction (during and after the first beat)?

- Do you hear an unusual sound during relaxation (during and after the second beat)?

3. Create a table in your science notebook to record your observations of each heart sound. You will need to record the number of each sound, the kind of sound you hear, and what the heart problem might be.

4. Listen to each abnormal heart sound again. Remember, you will first hear a normal heart sound. This will be followed by the abnormal heart sound. Describe each abnormal heart sound in your table.

5. Think about the types of heart problems that may occur, such as valve problems, a hole in the heart wall, or an irregular heartbeat. Try to identify the type of heart problem that may have caused each abnormal heart sound you heard.

ANALYSIS

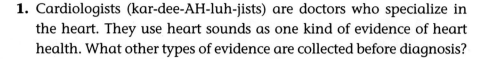

1. Cardiologists (kar-dee-AH-luh-jists) are doctors who specialize in the heart. They use heart sounds as one kind of evidence of heart health. What other types of evidence are collected before diagnosis?

2. Why are knowledge of heart valves and the sounds they make important to doctors? Explain.

3. Why do you think the doctor places the stethoscope on several different areas of your chest to listen to your heart beat?

4. Why might someone who has a heart defect such as weak heart valves or a small hole in the heart wall become breathless after climbing a long flight of stairs? Explain.

LABORATORY

Do you know someone who is concerned about his or her blood pressure? Maybe this person takes medicine or follows a special diet to help control blood pressure. Some people use blood pressure cuffs and stethoscopes to measure their blood pressure. They often keep careful records of their blood pressure from day to day.

What is high blood pressure? How does it affect the ability of your heart to pump blood?

CHALLENGE

What are the effects of high blood pressure on the heart?

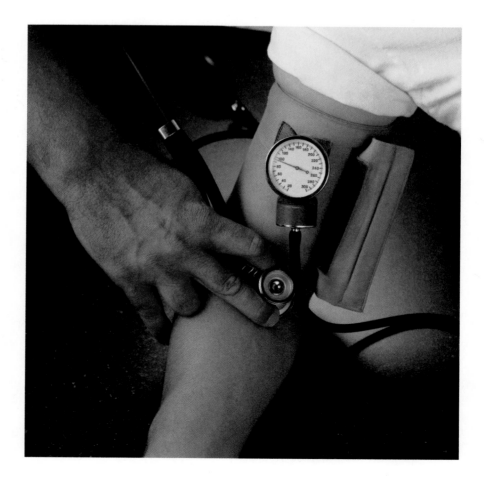

MATERIALS

For each group of four students
- 1 pressure bulb
- 2 45-cm (18-inch) lengths of clear plastic tubing
- 1 plastic bin (or bucket)
- 1 clamp
- 1 1-liter bottle or other measuring container
    access to a wall clock or watch with a second hand
    supply of water

For each student
- 1 Student Sheet 27.1, "Blood Pressure Data"

PROCEDURE

1. Fill the plastic bin approximately ¾ full of water. Your teacher may have already done this.

2. Place one end of the tubing in the water and establish the flow of water.

3. Each person in your group should perform one of the following roles: (Each member of your group will have a chance to perform each role.)

   a. Vein: Hold the tubing in the plastic bin of water. Be sure that the end of the tubing is in the water at all times!

   b. Artery: Hold the other end of the tubing in the 1-liter bottle and observe what happens to the water flow at different times while running the model. At the end of the trial, measure the volume of water pumped.

   c. Timer: Tell the group when to start and when to stop. Each trial should be 15 seconds long.

   d. Heart: Test your ability to pump the water by completing Steps 4–8.

➤

4. In 15 seconds, use one hand to pump as much water as possible from the plastic bin into the bottle.

5. Record the volume of water and your qualitative observations of the effort required to pump the water on Student Sheet 27.1, "Blood Pressure Data." (Use Table 1 for normal blood pressure. Use Table 2 for high blood pressure when using the clamp.)

6. Pour the water from the bottle back into the plastic bin.

7. Model high blood pressure by placing a clamp on the tubing. Slide the "artery" tubing (tubing leading to the bottle) through the clamp. Tighten the clamp by snapping it down three spaces, or clicks, to the third notch. (See Figure 1.)

8. Repeat Steps 4–6.

9. Rotate roles within your group. Repeat until all members of your group have completed Steps 4–8.

10. On Student Sheet 27.1, calculate the average volume of water the heart was able to pump under normal and high blood pressure.

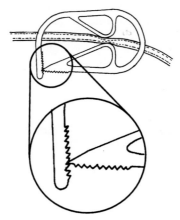

Figure 1: Placing the Clamp

11. Create a bar graph of the average volume of water the heart was able to pump under normal and high blood pressure. Be sure to label your axes and title your graph.

ANALYSIS

1. Squeezing the pressure bulb modeled the pumping of the heart and the work the heart does when blood pressure is normal.

    a. Recall what you observed when you acted as the artery. What happened to the flow of blood when the artery was clamped?

    b. How did clamping the artery affect how hard the heart had to pump?

2. What effect did clamping the artery have on the pressure of the water inside the tubing? What does it mean to have high blood pressure?

3. Explain what happens to the heart when you have high blood pressure as modeled in this activity. Support your answer with qualitative and quantitative data.

The tissues of your body need a constant supply of oxygen. Remember the circulation model from Activity 18, "The Circulation Game"? Imagine what would have happened if the flow of blood had stopped: the body would not get the oxygen it needs. The body's tissues can survive for only about 4 minutes without a fresh supply of oxygen. Two organs that are especially sensitive to a shortage of oxygen are the brain and the heart itself.

CHALLENGE

What causes a heart attack or a stroke?

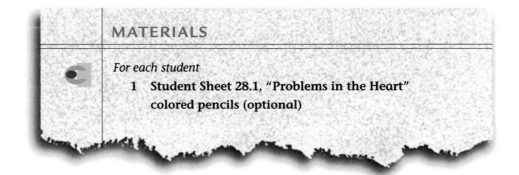

MATERIALS

For each student
1 Student Sheet 28.1, "Problems in the Heart"
   colored pencils (optional)

READING

The Heart Needs Blood Too

Blood goes through your heart as it is pumped. But the blood passing through the chambers of your heart does not supply oxygen to your heart muscle. Some of the blood that leaves your left ventricle travels back to your heart through arteries known as **coronary** (KOR-uh-nair-ee) **arteries**. The coronary arteries lead to a network of capillaries in your heart muscle. These capillaries carry the blood that provides oxygen to your heart muscle.

➢

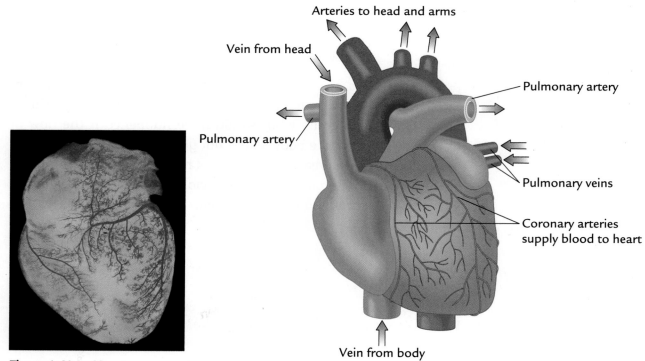

Figure 1: Your Heart

Arteries to head and arms

Vein from head

Pulmonary artery

Pulmonary artery

Pulmonary veins

Coronary arteries
supply blood to heart

Vein from body

If the blood supply is blocked or reduced, heart muscle can be damaged and pain results. This is commonly known as a heart attack. If a large enough part of the heart muscle is affected, then the heart can no longer pump blood. Heart failure and death result.

STOPPING TO THINK 1

a. Look at Stopping to Think 1 on Student Sheet 28.1, "Problems in the Heart." Think about what would happen if the blood flow were blocked at Point A on the diagram. On Student Sheet 28.1, shade the area of heart muscle that would be affected.

b. Think about what would happen if the blood flow were blocked at Point B on the diagram. On Student Sheet 28.1, shade the area of heart muscle that would be affected.

c. Which heart attack would be most likely to kill a person: the one caused by a blockage at Point A or the one caused by a blockage at Point B? Explain.

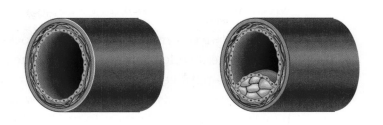

Figure 2: Normal and Blocked Arteries

How Heart Attacks Happen

Arteries supplying blood to the heart can become blocked by blood clots or by fat deposits. Look at Figure 2. In one condition called atherosclerosis (a-thuh-row-skluh-ROW-suss), deposits of fat partially block the center of the arteries. A diet high in some kinds of fat is associated with a higher risk of this condition. People who have high levels of this fat, called cholesterol, in their blood may modify their diet to reduce their cholesterol level.

STOPPING TO THINK 2

a. Look at Stopping to Think 2 on Student Sheet 28.1. In Artery A, draw a fat deposit that would block the artery just a little. In Artery B, draw a fat deposit that would block over half of the blood flowing through the artery.

b. Which of the two arteries (A or B) would be more likely to become completely blocked if a small blood clot came through the artery?

c. Why are people who are recovering from heart attacks sometimes given medicine to reduce the clotting activity of their blood?

Medical Help for Blocked Arteries

There are two common treatments for blocked arteries in the heart. In one treatment called coronary bypass, the surgeon inserts a segment of a vein—removed from another part of the body—around the blocked artery. This lets the blood flow through another pathway, bypassing the blocked artery. In another treatment called angioplasty [AN-jee-uh-plas-tee]), the surgeon inserts a tube with a balloon-like device into the partially blocked artery. The balloon is then inflated to stretch the artery. After the balloon is deflated and removed, the artery will be more open than before.

STOPPING TO THINK 3

Look at Stopping to Think 3 on Student Sheet 28.1. Imagine that tests reveal that a person has a blockage developing in the artery at Point X on Figures 4 and 5. Doctors are planning to do a coronary bypass operation. Does Figure 4 or Figure 5 show a bypass that will help this person's problem?

High Blood Pressure and Heart Disease

As the strongest muscle in your body, your heart first pumps blood through your arteries. This pumping action creates pressure within your blood vessels. As you discovered in Activity 27, "The Pressure's On," high blood pressure forces the heart to work harder to pump blood through the circulatory system. This can cause the heart muscle to become enlarged and weakened. It also strains the arteries, causing them to lose their ability to stretch. They become harder and less elastic. This is known as "hardening of the arteries." This process occurs with age, but it is more common in people with high blood pressure. With damaged arteries, blood clots or deposits of fat are more likely to form. When the arteries that supply the heart are damaged, a heart attack can follow. (A similar type of attack, known as a stroke, can happen if the arteries to the brain are damaged.)

In most cases, researchers do not know exactly what causes high blood pressure. Several factors, though, seem to increase a person's chance of having high blood pressure. Factors that increase the chance of an illness are called **risk factors**. Risk factors for high blood pressure include the person's age, weight, race, gender, and body type; whether the person smokes cigarettes; how much alcohol the person drinks; the person's diet and amount and frequency of exercise; and heredity (whether the person's parents or grandparents have high blood pressure). Older people, people with a family history of high blood pressure, people of African descent, and men in general have a higher risk of high blood pressure and other types of heart disease than other groups. Being extremely overweight, living an inactive life style, drinking large amounts of alcohol, and, in some people, eating a diet high in salt are associated with a greater risk of high blood pressure.

Smoking and Heart Disease

Cigarette smoking increases the risk of death from heart disease because of its impact on several risk factors. Smoking can increase blood pressure, increase the formation of fat deposits in the arteries, and increase the chance of blood clot formation in the arteries. A smoker is also less likely to survive a heart attack. The smoker's lungs are not as efficient in absorbing oxygen from the air and thus provide less oxygen to maintain the heart.

ANALYSIS

1. What kinds of health problems can be caused by blockages in coronary arteries?

2. Why should people with many risk factors for heart disease first check with a doctor before beginning an exercise program?

3. **Reflection:** What can you do to maintain or improve the health of your heart?

PROJECT

Having healthy habits now can reduce your risk of heart disease later. What factors increase your risk of heart disease? Which of these risks can you reduce? Find out by identifying which heart disease risks are **voluntary**—that is, under your control.

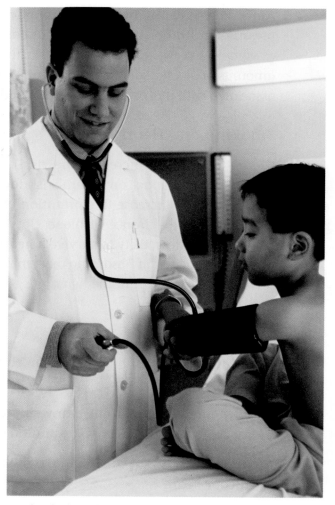

Regular check-ups help people find out if they have high blood pressure or other risks for heart or circulatory disease.

CHALLENGE

What is your relative risk of heart disease? How can you convince people to make choices that reduce their level of risk?

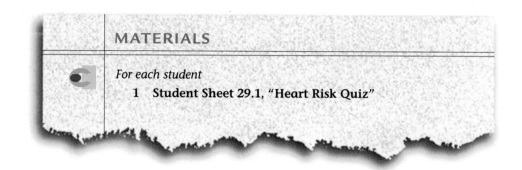

MATERIALS

For each student
1   **Student Sheet 29.1, "Heart Risk Quiz"**

PROCEDURE

Part One: Heart Risk Quiz

1. Evaluate your personal risk for heart disease by taking the quiz on Student Sheet 29.1, "Heart Risk Quiz." This quiz should be completed at home so you can ask questions of your family members. You will not have to share the results in class. Use the Scoring Guide on the back of the sheet to add up your risk points.

2. Look again at Student Sheet 29.1, "Heart Risk Quiz." Note that each item on the quiz is either about a risk (such as eating high-fat meals) or about a way to reduce a risk (such as eating fruits, beans, and vegetables).

   Decide with your group which risks are voluntary. If you think a risk is only partially voluntary, discuss why. Try to come to agreement within your group.

   • Listen to and consider the explanations and ideas of other members of your group.

   • If you disagree about a risk with others in your group, explain to the rest of the group why you disagree.

3. Many people know the risks for heart disease, but not all people work hard to reduce their risk. Look at the voluntary risks you identified in Step 2. Discuss with your group the possible reasons why young people like yourself may not change their habits.

Part Two: Helping Others Make Healthy Decisions

4. Design and produce a brochure persuading teenagers to make decisions to reduce their risk for heart disease.

Your brochure should do the following:

- Be clear and convincing.

- Describe one voluntary risk factor for heart disease.

- Explain how the circulatory system works.

- Explain how the risk factor you have identified affects the structure and function of the circulatory system.

- Suggest practical steps teens can take in their daily lives to reduce this risk factor.

- Use informal diagrams and artwork to help present the content of the brochure.

ANALYSIS

1. Imagine you are a doctor and Mr. Jacobs visits you to discuss his health. He says, "I feel great. I run 2 miles every day, lift weights, and eat a healthy diet. No red meat and fast food for me! I looked at the risk factors for heart disease and it looks like I have zero risk. I'm sure I'll never have any problems!"

   What would you tell Mr. Jacobs about his risk for heart disease? Support your answer with evidence.

2. Imagine you are a doctor and Ms. McDonald visits you to discuss her health. She says, "I just took a look at risk factors for heart disease. I fall into a lot of high-risk categories. I work long hours and don't have much spare time. I eat a lot of fast food hamburgers and fries, and I exercise only on Saturdays. Does this mean that I will definitely have heart disease?"

   What would you tell Ms. McDonald about her risk for heart disease? Support your answer with evidence and discuss any trade-offs involved in your recommendation.

Hint: To write a complete answer, first state your opinion. Provide two or more pieces of evidence that support your opinion. Then consider all sides of the issue and identify the trade-offs of your decision.

EXTENSION

Visit the SALI page of the SEPUP website for links to more information about heart disease risk. Do you think the materials you find there are successful at convincing readers to reduce their risk?

Micro-Life

C

Unit C

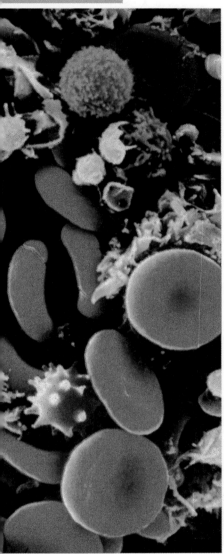

It was summer and Alex was spending two weeks at camp. In the middle of the first week, he discovered small spots all over his skin. The spots were so itchy that he went to the camp nurse.

"This rash I've got is really bothering me. Could I have some lotion?" Alex asked.

The nurse looked Alex over carefully. "I'm afraid you're going to need more than a little lotion. You've got the chicken pox!"

"What do you mean? Is that bad? I do feel a little sick, but I thought it was from the camp food," grinned Alex.

"Don't worry, you'll be fine," reassured the nurse. "But I know from the health records that there are people at camp who have never had the chicken pox nor have been vaccinated against it. So I'm going to have to isolate you in the sick room until your mom comes to take you home."

"Can my new friends visit me, at least?"

"I'm afraid not. To be safe, only the camp staff members we know have had the chicken pox will be allowed near you."

• • •

Is this fair? Should someone with a disease be isolated? Why were some people allowed to visit Alex while others were not? What information did the camp director have about this disease that helped her make this decision?

In this unit, you will explore what causes infectious diseases, how these diseases are transmitted, and how medicine is used to combat them.

MODELING

Different diseases are caused by different factors, such as germs, heredity, or even the environment. Some diseases caused by germs are **infectious** (in-FEK-shuss), which means that they can be passed from one person to another. Many infectious diseases, such as chickenpox, are more common among children. How quickly can an infectious disease spread among a group of people? What can be done to stop more people from getting sick?

CHALLENGE

How does an infectious disease spread in a community?

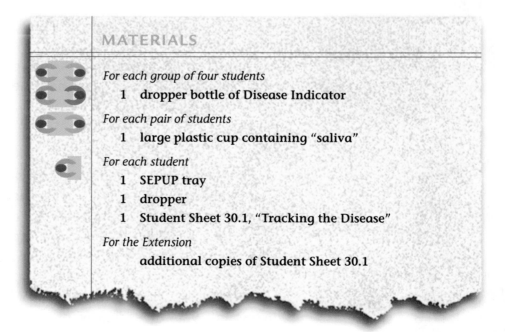

MATERIALS

For each group of four students
  1 dropper bottle of Disease Indicator

For each pair of students
  1 large plastic cup containing "saliva"

For each student
  1 SEPUP tray
  1 dropper
  1 Student Sheet 30.1, "Tracking the Disease"

For the Extension
  additional copies of Student Sheet 30.1

PROCEDURE

Part One: Planning Your Day

1. In Table 1, "My Movements" (on Student Sheet 30.1, "Tracking the Disease"), fill in the Place column by listing the one place or event that you plan to go to each day.

REMINDER

Good laboratory procedure means no accidental contamination! Be sure to follow the directions and be careful with your dropper.

2. Use your dropper to put 10 drops of "saliva" from the large plastic cup into large Cup A of your SEPUP tray.

3. Use your dropper to fill Cup B ¾–full of "saliva" from the large plastic cup.

4. After you and your partner have completed Steps 2 and 3, return the large plastic cup to your teacher.

Part Two: Meeting Other People

Your teacher will guide you through Steps 5–8.

5. On Day 1, go to the place you chose and recorded in Table 1 on Student Sheet 30.1. Then:

   a. Read the card to determine the number of people with whom you should exchange "saliva." (If no one else is at this place, you do not need to do anything.)

   b. Exchange "saliva" with people at this place by using your dropper to transfer 10 drops of solution from your Cup B into Cup B of the other student's tray while the other student transfers 10 drops of solution from his or her Cup B into Cup B of your tray. Cup B should now contain about the same amount of solution with which you started.

   c. Use your dropper to remove half of the solution from your Cup B and place it into Cup C of your own SEPUP tray.

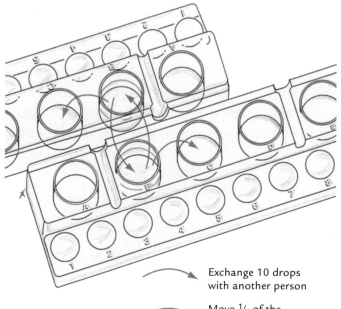

Exchange 10 drops with another person

Move ½ of the solution to Cup C

6. On Day 2, go to the place you chose in Table 1. Repeat Steps 5a and 5b, but this time, transfer solutions in Cup C.

7. Use your dropper to remove half of the solution from your Cup C and place it into Cup D of your own SEPUP tray.

8. On Day 3, go to the place you chose in Table 1. Repeat Steps 5a and 5b, but this time, transfer solutions in Cup D.

Part Three: Getting Tested

9. *Did you catch the disease?* Find out by testing Day 3 (Cup D) by adding 2 drops of Disease Indicator.

   If you have been infected with the disease, the solution will change color. If the solution does not change color, congratulations—you have escaped catching the disease this time! Record your results in Table 1 on Student Sheet 30.1.

10. *If you were infected with the disease, when did you get it?* Find out by testing your initial "saliva" (Cup A), Day 1 (Cup B), and Day 2 (Cup C). Record your results in Table 1 on Student Sheet 30.1.

11. Use the class data to complete Table 2, "Class Results," on Student Sheet 30.1.

12. Use the data in Table 2 to create a line graph of the number of infected people over time. Be sure to include the initial data (Day 0), to label your axes, and to title your graph.

EXTENSION

As a class, repeat the activity. Be sure to choose different places to visit or events to attend. Did the disease spread within your community in the same way? What similarities or differences do you observe? What role do the initially infected people play in affecting the spread of disease?

ANALYSIS

1. Use your graph of the class results to answer the following questions:

   a. What happened to the number of people infected with the disease over time?

   b. How does this compare to your initial prediction? Explain.

2. Think about how the infectious disease was spread from person to person in your community. If you were trying to avoid catching the disease, what could you do? Use evidence from this activity to support your answer.

3. **a.** Imagine that you are the director of the health department in the town where this disease is spreading. It is your job to help prevent people from getting sick. Explain what you would recommend to try to prevent more people from getting infected.

   b. What are the trade-offs of your recommendations?

4. What are the strengths and weaknesses of this model for the spread of infectious diseases?

5. Could you use this activity to model how diseases that are *not* infectious are spread? Explain.

PROJECT

You walk down the street and see a billboard that warns you about the risks of young people smoking. You read a magazine ad that tells you to drink more milk. You turn on the TV and watch a 10-second spot encouraging you to read more. These ads, which aren't selling a specific brand or service, are known as public service announcements (PSAs). PSAs provide useful and important information to the public. Many PSAs encourage children and younger adults to make choices to ensure long-term health. People who put out PSAs are responsible for making sure that the information is accurate and helpful. This means that claims should be supported by research or scientific studies. In this activity, you will make a PSA about a disease.

Disease is simply a breakdown in the structure or function of a living organism. In humans, there are many different ways that our structures (such as tissues and organs) and our functions (such as digestion) can be affected. As a result, many different diseases can affect people. Before you make your PSA, you will need to learn about a disease and decide what information is important to share with others.

"In 14 years of modeling, this is my favorite shot of myself."

Christy Turlington considers quitting smoking her biggest success One of her biggest regrets was that she ever started.

CDC
CENTERS FOR DISEASE CONTROL
AND PREVENTION

Many PSAs encourage younger adults and children to make choices that ensure long-term health.

CHALLENGE

What type of information should be presented in a PSA on a disease?

MATERIALS

For the class

    books, magazines, CD-ROMs, Internet access, etc.

For each student

1   Student Sheet 31.1, "The Hunt Is On"

1   Student Sheet 31.2, "Knowing About Disease" (optional)

1   Student Sheet 31.3, "Disease Research Report"

1   Student Sheet 31.4, "PSA Panels" (optional)

PROCEDURE

Part One: Knowledge of Disease

1. Use Student Sheet 31.1, "The Hunt Is On," to find people in your class who know someone who has had a particular disease. Have them initial the box with the name of the disease. Each person (including you) can initial only one box on your Student Sheet.

2. On Student Sheet 31.2, "Knowing About Disease," mark an "X" in the "Student" column if you know someone who has had that disease.

3. Use Student Sheet 31.2 to find out whether a parent/guardian and/or a grandparent/older adult have known someone with a particular disease.

4. As a class, total the number of students, parents/guardians, and/or grandparents/older adults who have known someone with a particular disease.

Part Two: Disease Research

5. Choose a disease to research. You may want to choose from the list of diseases on Student Sheet 31.2.

6. Over the next few days or weeks, find information on this disease from books, magazines, CD-ROMs, the Internet, and/or interviews. You can also go to the SEPUP website to link to sites with more information on diseases mentioned in this activity.

7. Use the information you find to complete Student Sheet 31.3, "Disease Research Report." You should be able to describe

   a. what causes this disease

   b. symptoms of this disease

   c. how this disease is spread among humans

   d. how this disease can be prevented

   e. how this disease is medically treated

   f. two important and/or interesting facts about this disease.

8. Use your Disease Research Report to create a public service announcement (PSA) in the form of a cartoon strip. Develop a 3–6 panel cartoon strip that tells people either how to prevent getting the disease or at least one important piece of information about the disease. An example is shown below. Remember, you can use humor, but be sure your PSA is appropriate for the classroom!

ANALYSIS

Part One: Knowledge of Disease

1. For which diseases was it easy to find someone to initial your boxes on Student Sheet 31.1?

2. Would you expect to find that the same diseases are equally common in different parts of the world? Why or why not?

3. Compare the number of students, parents, and grandparents who knew someone with a particular disease. What patterns do you observe? For example, which diseases were more familiar to the grandparent generation than your generation? What do you think is the reason for this?

Part Two: Disease Research

Look at the PSAs produced by other students.

4. What can you do to prevent catching an infectious disease?

5. What types of diseases cannot be prevented? Explain.

INVESTIGATION

Epidemiologists (eh-puh-dee-mee-AH-luh-jists) are scientists who trace the spread of a disease through a population. They do this to learn how the disease spreads and to find ways to help prevent its further spread. One way epidemiologists gather such information is by going to a community and comparing sick people with healthy people. Their work is complicated because not all people who are exposed to an infection get sick. Vaccinations, previous exposure to the infection, and overall health affect whether a person who is exposed to an infection will become sick. Sometimes there is no obvious connection among the sick individuals. Epidemiologists then have to resort to testing for the infection in healthy people to find the transmission path. This is because you can pass on an infection before you know you are sick. Some infectious diseases such as typhoid (TIE-foyd) and diphtheria (dip-THEER-ee-uh) can be carried (and spread) for a long time by someone who never develops symptoms of the illness. These **carriers** can be important links in the spread of a disease.

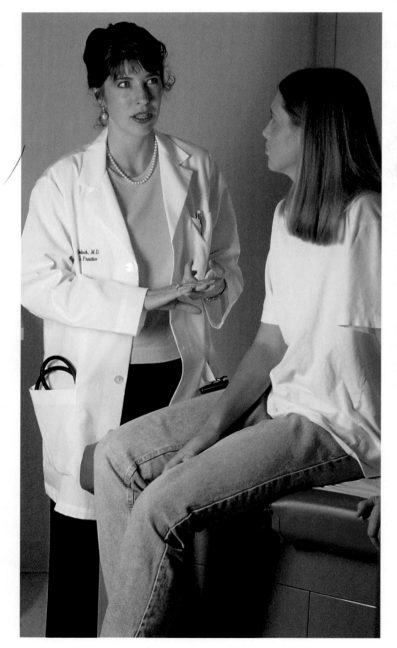

The Abingdon Chronicle

December 15

New Disease at Salk Junior High School?

Yesterday, a science teacher at Salk Junior High School was sent to the hospital with a serious illness. Lab results came back with a surprising diagnosis of a new type of infection. Why the surprise? Because this type of infection has never been observed before in Abingdon.

Ms. Shah became sick on December 3 with a high fever, sore throat, and wheezing cough. At first she thought she had the flu, but when she had difficulty breathing, she went to her doctor. Dr. Holmes of Abingdon Hospital noticed similar symptoms in another patient who is a student at Salk. When interviewed, he commented, "Ms. Shah's symptoms were very serious. She

was sick for a week and in the hospital for two days. Because Ms. Shah comes into contact with many students each day, I was concerned that this disease might spread. However, there were no more cases for about 10 days. I was so relieved!"

Since then, however, several more people at the school have become sick and have had similar symptoms. Local health officials are concerned. Could this be the beginning of an outbreak? A local epidemiologist, Dr. Montagu, is working to identify the path of transmission: who infected whom? In order to determine this, she has begun to gather information. To date, she has interviewed eight individuals.

CHALLENGE ➡

Who is (or are) the carrier(s) of the disease?

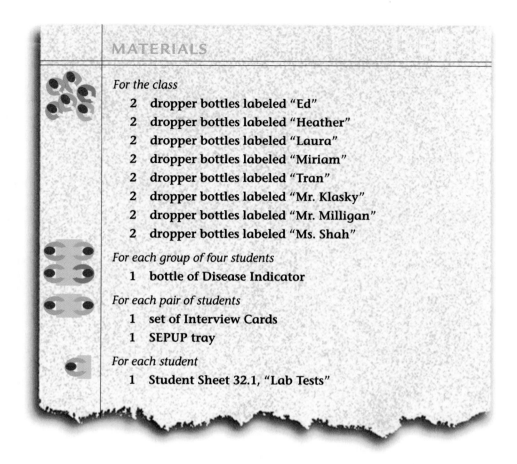

MATERIALS

For the class

2	dropper bottles labeled "Ed"
2	dropper bottles labeled "Heather"
2	dropper bottles labeled "Laura"
2	dropper bottles labeled "Miriam"
2	dropper bottles labeled "Tran"
2	dropper bottles labeled "Mr. Klasky"
2	dropper bottles labeled "Mr. Milligan"
2	dropper bottles labeled "Ms. Shah"

For each group of four students

1	bottle of Disease Indicator

For each pair of students

1	set of Interview Cards
1	SEPUP tray

For each student

1	Student Sheet 32.1, "Lab Tests"

PROCEDURE

Part One: Evidence from Interviews

1. Read the information on the Interview Cards.

2. Discuss with your partner how the disease may have spread from person to person.

3. Move the cards around to develop a web showing who could have caught the disease from whom.

4. In your science notebook, draw what you think is the web of disease transmission. Be sure to include how you think these people are connected. For example:

Ms. Shah (symptoms) $\xrightarrow{\text{science class}}$ Heather (symptoms)

5. Discuss with your partner which people may be carriers of the disease. Record what you think in your science notebook.

6. In order to test your hypothesis, you will be able to test samples of "saliva" from these people for the presence of the disease. Based on your hypothesis, record on Student Sheet 32.1, "Lab Tests," the names of four people you would most like to test.

Part Two: Collecting Lab Evidence

REMINDER

Good laboratory procedure means no accidental contamination! When using a dropper bottle, unscrew the lid but do not put the lid down on the table. Instead, use the bottle and immediately re-cap it.

7. Find the dropper bottle for one of the people you would like to test. It contains that person's "saliva." Place 3 drops of the "saliva" sample into one of the small cups in your SEPUP tray.

8. Test the sample by adding 2 drops of Disease Indicator. Make sure the dropper does not touch the "saliva." A positive test for the disease will show a pink color.

9. Record your result on Student Sheet 32.1.

10. Repeat Steps 7–9 for the rest of the people you are testing.

ANALYSIS

1. Based on your test results, draw a web showing your proposed path of disease spread. In your web, identify who is infected, the dates that he or she became sick, and whether the person is a carrier. How does this web compare to your original hypothesis?

2. a. Who was (or were) the carrier(s) of the disease?

b. What evidence do you have to support your answer?

3. Think back to the suggestions you made to prevent the spread of disease when discussing Analysis Questions 2 and 3 of Activity 30, "It's Catching!" How does the knowledge that some diseases can be spread by carriers affect your ideas? In other words, what recommendations would you make to a community that was experiencing a disease outbreak?

4. A group of Abingdon parents have demanded that the family members and close friends of all infected individuals, including students and teachers, stay home until everyone with symptoms gets better. Explain whether you agree with their demand. Support your answer with evidence and identify the trade-offs of your decision.

Hint: To write a complete answer, first state your opinion. Provide two or more pieces of evidence that support your opinion. Then consider all sides of the issue and identify the trade-offs of your decision.

5. Reflection: Explain whether you would change your answer to Question 4 if the disease had more severe symptoms and a greater chance of causing death.

VIEW AND REFLECT

Humans are not the only organisms that can spread disease. Some diseases, such as the bubonic (byu-BAH-nick) plague (PLAIG) and malaria (muh-LAIR-ee-uh), are spread by vectors (VEK-terz). A **vector** is an organism (other than a person) that spreads disease-causing germs usually without getting sick itself. Rats, ticks, mosquitoes, and fleas can act as vectors for various human diseases. Ticks, for example, spread Lyme disease. That's why it's important to wear long sleeves and pants to avoid picking up ticks when hiking in some areas.

CHALLENGE ➡

What is the role of vectors in spreading disease?

PROCEDURE

1. In order to prepare to watch the story on the video, first read Analysis Questions 1–3.

2. Watch a segment on the bubonic plague from the video, *A Science Odyssey*: "Matters of Life and Death."

ANALYSIS

1. In 1900, people did not know how the bubonic plague was spread. What did officials do to try to stop the spread of disease?

2. a. Draw a diagram showing how the bubonic plague is spread.

b. Identify the vector of this disease.

3. By 1906, officials knew how the bubonic plague was spread. What did they do this time to stop the spread of disease?

4. a. Malaria, a disease particularly common in Africa, is caused by a tiny germ known as *Plasmodium*. When a female mosquito bites a person infected with malaria, she sucks up *Plasmodium* along with the blood. When she bites a healthy person, germs in her saliva infect that person. What is the vector in this case?

b. Now that you know the vector of malaria, suggest two ways that the spread of malaria could be reduced or prevented.

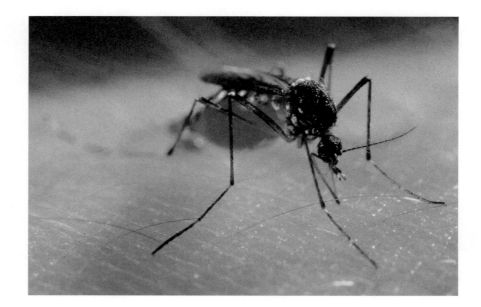

TALKING IT OVER

When the bubonic plague first occurred in San Francisco in 1900, the official response was to isolate, or quarantine (KWOR-un-teen), Chinatown. By 1906, it was clear that quarantine would not work to stop a disease spread by fleas. But during that first outbreak, people living in Chinatown were left to fend for themselves.

Leprosy (LEH-pruh-see), or Hansen's disease, is another illness that caused people to be quarantined. Because this is a long-term disease, people infected were forced to live apart from others for many years. In this activity, you will read about leprosy and make decisions about how people with leprosy and other infectious diseases should be treated.

A doctor checks the skin of a young boy receiving drug treatment to fight leprosy.

CHALLENGE

How should people with infectious diseases be treated?

PROCEDURE

1. Read the story of leprosy. As you read, think about how infectious diseases can and should be controlled.

2. Discuss Analysis Questions 1 and 2 with your group.

THE STORY OF LEPROSY

Imagine having a disease that, if people knew you had it, would cause you to be taken away from your family and forced to live somewhere else. If you had this disease, other people would be afraid to come near you or touch you. You would not be allowed to eat or sleep near uninfected people.

This is the story of leprosy. Historically, people with leprosy have been expelled from society. In the Middle Ages, people who had leprosy were considered dead. They were given a funeral service while still alive, and then forced to wander, without a home, and beg for food until they died. People have always been afraid of those with leprosy because the disease could cause serious deformities of the face, arms, and legs. Nerve damage could cause skin numbness. People sometimes lost fingers and toes to injuries that they did not feel. Damage to optic nerves could lead to blindness. In addition, there were often sores on the skin. Many of these changes were permanent and left those infected physically disabled.

In 1894, the Louisiana Leper Home was established in Carville, Louisiana. At that time, people infected with leprosy were not allowed on public transportation and were taken to the Home by boat. In fact, some individuals were placed in handcuffs and leg shackles so they would not escape. Once there, they were not allowed to make telephone calls or vote. Fear of spreading the disease meant that their outgoing mail and money were sterilized. The local soft drink company would not even accept empty bottles from the Home.

The germ that causes Hansen's disease can be found in certain animals, including armadillos (shown below). Could armadillos be vectors for this disease? Scientists don't yet know the answer.

In 1873, Dr. G. A. Hansen discovered that the bacterium *Mycobacterium leprae* caused leprosy. However, no one knew how to prevent the bacteria from spreading. It was not until the late 1950s that the use of antibiotics against leprosy finally allowed people infected with the disease in the United States to choose where they wanted to live. Today, leprosy is called Hansen's disease in honor of Dr. Hansen. This new name also reflects the modern attitude toward this disease. It is now possible to treat and cure Hansen's disease with drugs. An infected person can become non-contagious after just a few days of treatment, and the spread of the disease can be controlled without long-term isolation of the victims.

The irony is that Hansen's disease does not spread easily and is very hard to catch. Only about 5% of family members living with infected people develop the disease. The exact way this disease spreads is still not known. Scientists believe that becoming infected requires close contact with an infected person over a long period of time. However, it has always been rare even for people caring for those with Hansen's disease to catch it. This may be because more than 90% of the population is believed to be immune; this means that these people would not become infected even after being exposed to the disease.

Although it is now rare in the United States, Hansen's disease is still a serious health problem in parts of Asia, Africa, and South America (particularly Brazil), where it usually affects the poorest people. Despite advances in its treatment, more than one new case of Hansen's disease is diagnosed worldwide each minute.

EXTENSION

Go to the SALI page of the SEPUP website to link to sites with more information about the history of leprosy in the United States.

ANALYSIS

1. How have people with Hansen's disease been treated throughout history? Provide specific examples.

2. Imagine that you meet someone who tells you that he or she has Hansen's disease. How would you respond? Support your answer with evidence from the activity.

3. Discuss what factors should determine how a person with an infectious disease should be treated.

4. Based on your understanding of infectious diseases, explain whether you think people who have an infectious disease should be quarantined. Support your answer with evidence and identify the trade-offs of your decision.

   Hint: To write a complete answer, first state your opinion. Provide two or more pieces of evidence that support your opinion. Then consider all sides of the issue and identify the trade-offs of your decision.

5. **Reflection:** In Activities 30, 32, and 33, you considered how to prevent the spread of infectious disease. Imagine that you were infected with an infectious disease. Would you volunteer to be quarantined? Explain.

LABORATORY

Scientific discoveries often follow the development of new tools and technologies. This is certainly true in the case of infectious diseases. As you saw in Activity 33, "From One to Another," researchers Alexandre Yersin and Shibasaburo Kitasato independently used the microscope to identify the cause of the bubonic plague. Compound microscopes—microscopes that use more than one lens—were invented around 1595. These first microscopes usually magnified objects only 20–30 times their original size. But as you will learn in the next few activities, even this level of magnification was enough to discover a world of new scientific ideas.

By 1840, Italian physicist Giovanni Amici (a-MEE-chee) had invented the oil-immersion microscope which could magnify objects 6,000 times. In most middle schools, the highest level of magnification is usually about 400 times. Today, the transmission electron and scanning electron microscopes can magnify objects over 40,000 times!

CHALLENGE

What is the correct way to use a microscope?

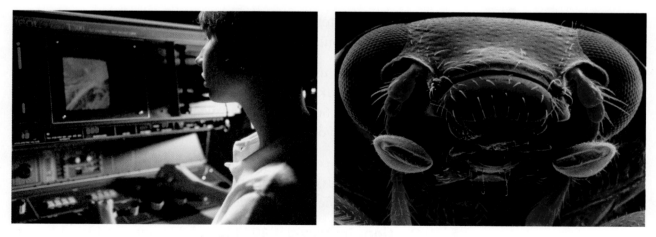

The photograph on the left shows a scanning electron microscope, which shows the surfaces of objects. In the photograph on the right, you can see the head of a ladybug as seen through this type of microscope magnified 23 times.

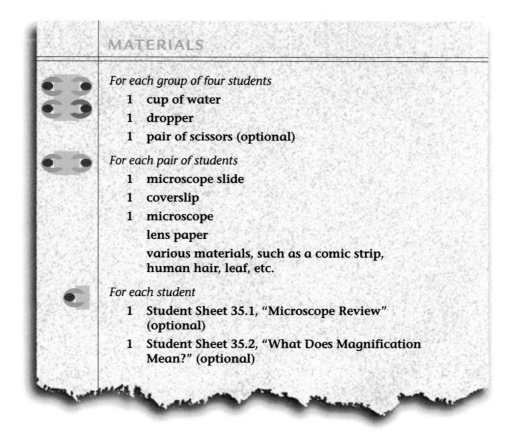

MATERIALS

For each group of four students
- **1 cup of water**
- **1 dropper**
- **1 pair of scissors (optional)**

For each pair of students
- **1 microscope slide**
- **1 coverslip**
- **1 microscope**
- **lens paper**
- **various materials, such as a comic strip, human hair, leaf, etc.**

For each student
- **1 Student Sheet 35.1, "Microscope Review" (optional)**
- **1 Student Sheet 35.2, "What Does Magnification Mean?" (optional)**

PROCEDURE

Part One: Earning a License

1. Your teacher will demonstrate the different parts of a microscope, as shown in Figure 1.

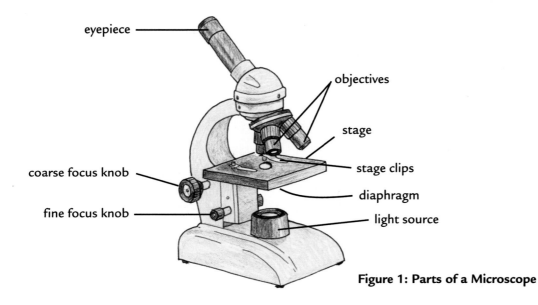

Figure 1: Parts of a Microscope

RULES FOR HANDLING A MICROSCOPE

- Always carry a microscope using two hands, as shown in the picture to the right.

- Rotate the objectives carefully. Do not allow them to touch the stage or anything placed on the stage, such as a slide. This can damage the microscope.

- When using the coarse focus knob, begin with the stage in its highest position and always focus by lowering the stage (away from the objective).

- Use only lens paper to clean the eyepiece or the objectives.

- When you have finished using a microscope, remember to turn off the microscope light and set the microscope back to low power (the shortest objective, usually 4x).

2. As a class, discuss the rules for handling a microscope.

3. Demonstrate your knowledge of the microscope, as required by your teacher.

4. Collect your microscope license! You are now ready to begin using a microscope.

Part Two: Using the Microscope

5. Clean your microscope slide and coverslip by rinsing them with water and gently wiping them dry.

6. With your partner, look at the materials list posted by your teacher and decide what you will examine under the microscope.

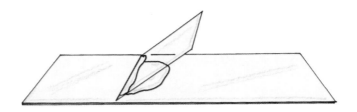

Figure 2: Placing the Coverslip

7. Place the material (or a small piece of the material) flat on the center of your microscope slide.

8. Use a dropper to place a drop of water directly onto your material. Carefully touch one edge of the coverslip to the water at an angle (as shown in Figure 2). Slowly allow the coverslip to drop into place.

9. Be sure that your microscope is set on the lowest power (shortest objective) before placing your slide onto the microscope stage. Center the slide so that the material is directly over the light opening and adjust the microscope settings as necessary.

   Hint: To check that you are focused on the material that is on the slide, move the slide slightly while you look through the eyepiece—the material that you are focused on should move at the same time you move the slide.

10. Begin by observing the sample with low power (usually the 4x objective). In your science notebook, describe how the material looks under low power compared to how it looks with your eyes. When you find an area you would like to explore at higher magnification, use the stage clips to secure the slide.

   Hint: If material on the slide is too bright to see, reduce the amount of light on the slide by slightly closing the diaphragm under the stage.

11. Without moving the slide, switch to medium power (usually the 10x objective). Adjust the microscope settings as necessary.

Hint: If material on the slide is too dark to see, increase the amount of light on the slide by slightly opening the diaphragm under the stage.

12. When you have finished using the microscope, turn off the microscope light and set the microscope back to low power (usually the 4x objective).

ANALYSIS

1. How does the microscope change the image you see? **Hint:** Compare the material you placed on the stage with what you see through the eyepiece.

2. Describe how the material(s) that you observed looked under low as compared to medium power. What differences did you observe? How did this compare to what you saw with your eyes?

3. The microscope is one important tool used by scientists to study living things. What other tools are used by life scientists? Think about tools used by doctors and in laboratory and field research. List three tools used by life scientists and describe the kind of information they can help scientists collect.

If someone asked you what makes you sick, you might answer that germs, bacteria, or viruses make you sick. During the early 1900s, some people thought an infectious disease like the flu could be caused by nakedness, contaminated food, irritating gases in the atmosphere, unclean clothing, open windows, closed windows, old books, dirt, dust, or supernatural causes.

What does cause infectious diseases? You can begin to answer this question with the study of **microbes** (MY-krobz), another word for creatures that are too small to be seen with the human eye. Some of these microbes cause diseases. In this activity, you will look for aquatic (water) microbes. You can see some examples of microbes in the photographs below.

CHALLENGE

What kinds of microbes can you find?

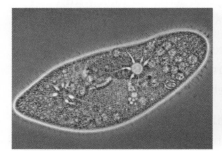

Paramecium

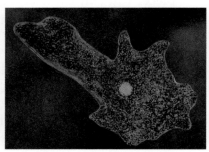

Amoeba

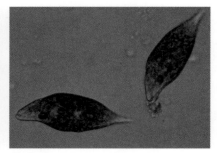

Euglena

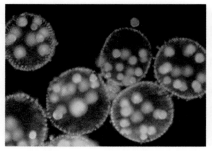

Green algae

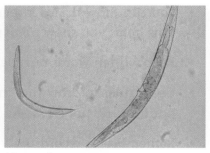

Nematodes

Stentor

MICROSCOPE DRAWING MADE EASY

Below is a picture taken through a microscope of the alga *Spirogyra*. The diagram to the right shows what a biologist or biological illustrator might draw and how he or she would label the drawing. Did you know that some artists draw only scientific illustrations?

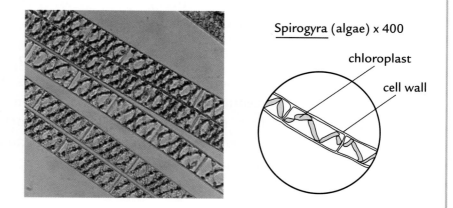

Spirogyra (algae) x 400

chloroplast

cell wall

Some tips for better drawings:

- Use a sharp pencil and have a good eraser handy.

- Try to relax your eyes when looking through the eyepiece. You can cover one eye or learn to look with both eyes open. Try not to squint.

- Look through your microscope at the same time as you do your drawing. Look through the microscope *more* than you look at your paper.

- Don't draw every small thing on your slide. Just concentrate on one or two of the most common or interesting things.

- You can draw things larger than you actually see them. This helps you show all of the details you see.

- Keep written words outside the circle.

- Use a ruler to draw the lines for your labels. Keep lines parallel—do not cross one line over another.

- Remember to record the level of magnification next to your drawing.

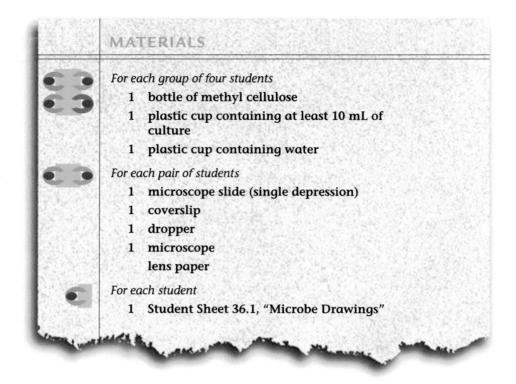

MATERIALS

For each group of four students
1   bottle of methyl cellulose
1   plastic cup containing at least 10 mL of culture
1   plastic cup containing water

For each pair of students
1   microscope slide (single depression)
1   coverslip
1   dropper
1   microscope
   lens paper

For each student
1   Student Sheet 36.1, "Microbe Drawings"

PROCEDURE

1. Clean your microscope slide and coverslip by rinsing them with water and gently wiping them dry.

2. Use the dropper in the cup containing culture to place a drop of liquid from that cup onto your slide.

3. After placing a drop of the culture liquid on the slide, add one drop of methyl cellulose directly on top of the first drop. Be careful not to add more than one drop! The methyl cellulose will slow down the movement of the microbes.

4. Carefully touch one edge of the coverslip, at an angle, to the liquid on your slide (as shown in Figure 1). Slowly allow the coverslip to drop into place.

Figure 1: Placing the Coverslip

5. Be sure that your microscope is set on the lowest power (shortest objective) before placing your slide onto the microscope. Center the slide so that the specimen is directly over the light opening and adjust the microscope settings as necessary.

➤

Hint: To check that you are focused on the material that is on the slide, move the slide slightly while you look through the eyepiece—the material that you are focused on should move at the same time you move the slide.

6. Begin by observing the sample on low power (usually the 4x objective). You may need to search the slide for signs of micro-life, or you may observe microbes moving through your field of view.

Hint: If material on the slide is too light to see, reduce the amount of light on the slide by slightly closing the diaphragm under the stage.

7. Without moving the slide (which can be secured with stage clips), switch to medium power (usually 10x). Adjust the microscope settings as you look again for signs of micro-life.

8. Without moving the slide, switch to high power (usually the 40x objective). *Be careful not to smash the objective against the slide!* Adjust the microscope settings as necessary. Search the slide for signs of micro-life. Some of the microbes may be very small, so look carefully!

Hint: If material on the slide is too dark to see, increase the amount of light on the slide: do this by slightly opening the diaphragm under the stage.

9. Review "Microscope Drawing Made Easy" on page C-28.

10. Either on medium or high power, draw at least two microbes.

11. When you have completed Step 10, turn off the microscope light and set the microscope back to low power (usually the 4x objective).

ANALYSIS

1. Is it possible that microbes exist that are smaller than those you observed? Explain how you might try to collect evidence to prove or disprove your idea.

2. As a scientist, you are asked to describe two of the microbes that you saw to someone who has never looked through a microscope. Write a short paragraph describing the microbes that you observed.

3. **Reflection:** Imagine that you are a researcher studying microbes. Would you choose to study a disease-causing microbe or one that does not cause disease? Explain.

Microbes, just like the ones you observed in Activity 36, "Looking for Signs of Micro-Life," were discovered in the late 1600s. But the idea that such tiny organisms could cause disease did not develop until the 1860s, less than 150 years ago. Why did it take so long to figure this out?

CHALLENGE

How did the germ theory of disease develop?

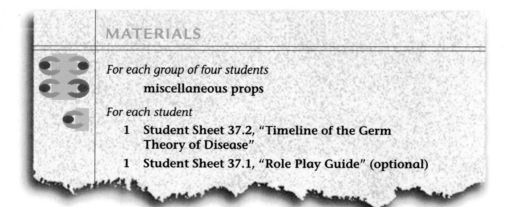

MATERIALS

For each group of four students
   miscellaneous props

For each student
   1   **Student Sheet 37.2, "Timeline of the Germ Theory of Disease"**
   1   **Student Sheet 37.1, "Role Play Guide" (optional)**

PROCEDURE

1. Your group will be assigned one of the sections under "Cast of Characters."

2. With your group, read the section carefully and identify the important contribution(s) to science made by your character(s).

3. Develop a skit to present these important points to your class. You can make your skit historical (for example, show how your scientist made the discovery) or modern (for example, create an ad to sell Hooke's book). Be sure to create a role for each person in your group. You can use Student Sheet 37.1, "Role Play Guide," to help you.

4. Collect any props or additional materials you require.

5. Present your skit to the class.

6. As you watch the various skits, complete the timeline on Student Sheet 37.2, "Timeline of the Germ Theory of Disease."

CAST OF CHARACTERS

Robert Hooke (1635–1703)

The late 17th century was a period of great scientific discovery. While many people offered theories without experimentation or evidence, English scientist Robert Hooke believed that good science resulted from making observations on what you could see. In his twenties, he wrote a book of his observations and drawings of the natural world called *Micrographia*, meaning "tiny drawings." It was first published in 1665. In this one book, he presented his ideas about the life cycle of mosquitoes, the origin of craters on the moon, and fossils. But Hooke is most remembered for including drawings of what he saw through a microscope.

Figure 1: Hooke's Microscope

Hooke developed his own version of the compound microscope (see Figure 1), and it was one of the best available at the time. Today, his most famous drawing from *Micrographia* is a drawing of cork—the same kind of cork that is used in corkboards and bottle stoppers. Since cork is made from the bark of the cork oak tree, it is essentially dead plant tissue. Using his microscope, Hooke looked at very thin slices of cork. He noticed what looked like little rooms (see Figure 2b). Because of this, he called these shapes **cells**, another word for *rooms*. With this simple observation, Hooke introduced an idea that would become the basis of new fields in biology—but not for almost 200 years!

Figure 2a: Cork Tree

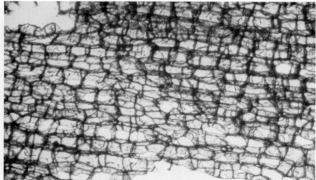
Figure 2b: Cork Cells

Anton van Leeuwenhoek (1632–1723)

Anton van Leeuwenhoek (LAY-vun-hook) was a cloth salesman in Holland and an amateur scientist. He knew how to make very simple microscopes. (Today they would be considered magnifying glasses.) But he did not become interested in studying the microscopic world until he read Hooke's *Micrographia* (see the section on Robert Hooke), which was a very popular book at the time.

Leeuwenhoek's skill at building microscopes (like the one in Figure 3) enabled him to magnify objects over 200 times. This, combined with his curiosity, led to observations almost identical to those that you made in Activity 36, "Looking for Signs of Micro-Life." In 1673, Leeuwenhoek described what he saw in a drop of water: "...wretched beasties. They stop, they stand still...and then turn themselves round...they [are] no bigger than a fine grain of sand." By examining scrapings from his teeth, he found additional evidence of these "many very little living animalcules, very prettily a-moving." Leeuwenhoek was one of the first people to observe and record microbes. He continued his observations until the end of his life.

Figure 3: Leeuwenhoek's Microscope

The small hole in the board contained the magnifying lens. The material to be observed was placed on the point in front of the lens.

HOW LEEUWENHOEK DESCRIBED *SPIROGYRA*

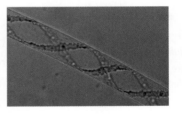

Look at the picture of the green alga *Spirogyra* shown above. Since Leeuwenhoek was not a good artist, he wrote very precise descriptions of his observations. (In addition, he hired someone to make drawings to go with his descriptions.) On September 7, 1674, he described *Spirogyra*, which can be found on lakes: "Passing just lately over this lake...and examining this water next day, I found floating...some green streaks, spirally wound serpent-wise. The whole circumference of each of these streaks was about the thickness of a hair of one's head...all consisted of very small green globules joined together: and there were very many small green globules as well."

How do Leeuwenhoek's descriptions of micro-life compare with your own?

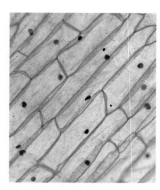

Figure 4: Plant Cells
Plants are made up of cells, like the plant cells shown in the photo above.

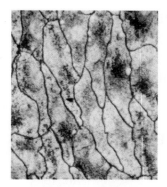

Figure 5: Animal Cells
Animals are made up of cells, like the skin cells shown in the photo above.

Figure 6: A Microbial Cell
Many microbes are made up of just one cell, like the microbe shown in the photo above.

Matthias Jakob Schleiden (1804–1881)
Theodor Schwann (1810–1882)
Karl Theodor Ernst von Siebold (1804–1885)

Over the next 150 years, scientists continued to use the microscope to study living organisms such as plants, insects, and microbes. But by the early 1800s, most botanists—scientists who study plants—were not using microscopes. They were busy naming and describing entire plants. German biologist Matthias Schleiden (SHLY-dun) was an exception. Although he was trained as a lawyer, he left the law to become a professor of botany. Schleiden preferred to use a microscope to study plants. (Look at Figure 4 to see what Schleiden may have seen.) Based on his study, he suggested in 1838 that all plants are made of cells. This was a completely new idea: just as a house could be made up entirely of bricks, plants were made up entirely of cells!

Schleiden knew another German biology professor, Theodor Schwann, who spent his time studying animals. Schwann was particularly interested in the digestive system. In 1839, one year after Schleiden proposed his theory, Schwann suggested that animals, and not just plants, were made up of cells. You can see animal cells in Figure 5. Because of their ideas, Schleiden and Schwann are credited with developing the **cell theory**: that all living organisms are made up of cells.

Other scientists began to build on Schleiden's and Schwann's ideas. In 1845, Karl Theodor Ernst von Siebold (SEE-bold) suggested that microbes were also made up of cells—or more specifically, one cell (see Figure 6). In fact, Siebold believed that organisms made up of many cells, like animals, were built out of single-celled microbes! While Siebold was wrong about this idea, he was right in stating that microbes were living creatures made up of the same material as animals and plants.

Rudolf Carl Virchow (1821–1902)

Why is Schleiden's and Schwann's cell theory important for understanding infectious disease? Their work influenced Rudolf Carl Virchow (VIR-koh), a Polish doctor. He had been treating and studying ill patients for many years. He is famous for saying, in the 1850s, "all cells arise from cells," meaning that cells reproduce to create new cells. He was right. When you see a new plant or a baby animal, you see a **multicellular** (many-celled) creature. All living organisms begin as a single cell. Most microbes are made up of *only* a single cell,

Did you know that each of these organisms is multicellular? Each is made up of millions of cells. You could collect evidence of this with a microscope.

as Siebold believed (see the section on Schleiden, Schwann, and Siebold). The cells of some living organisms, like people, continue to divide and grow. An adult human being is made up of about 10 trillion cells!

Virchow applied his ideas to disease. He knew that all cells grow from other cells. He thought that all diseases are caused by cells that do not work properly. He believed that diseased cells come from other usually healthy cells of the sick person. Virchow's ideas about disease were not completely correct, although they are correct for some diseases. His ideas were based on his work with leukemia (loo-KEE-mee-uh), which is a cancer of the blood. Cancer and other hereditary diseases are diseases of the cell. They *are* caused by cells that do not work properly. Infectious diseases are different, as scientists after Virchow discovered.

Ignaz Philipp Semmelweiss (1818–1865)

At the same time that Schleiden, Schwann, and Siebold were developing their ideas on cells, a Hungarian doctor working in Austria was trying to prevent young women from dying. It was the 1840s and pregnant women often died of a disease called childbed fever. Dr. Semmelweiss (sem-ul-VICE) noticed that many pregnant women were examined by

➢

doctors who had just completed an autopsy. He also observed another doctor die of childbed fever after he cut himself on the scalpel he was using to perform an autopsy. Semmelweiss concluded that childbed fever must be infectious and could be spread from something found in the dead bodies. He believed that doctors were carrying the disease from patient to patient.

Semmelweiss decided to try washing his hands between patients. As a result, fewer of his patients died. In two years, he reduced the death rate among his patients from 12% to 1%. He encouraged other doctors to use a strong chemical solution to wash their hands between patients. But because Semmelweiss could not explain why hand washing worked, many doctors refused to change their ways.

Semmelweiss tried hard to get hospitals to change their policies, but many people resisted his ideas. He eventually suffered a mental breakdown and died soon after. Within 10 years of his death, the development of the germ theory of disease would explain what he could not—that hand washing reduces the risk of infectious disease by removing germs like the ones shown in Figure 7.

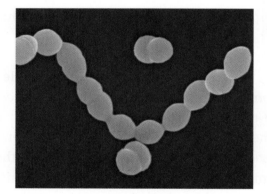

Figure 7
Today, childbed fever is called puerperal infection. It is caused by several different microbes, including Streptococcus (similar to the microbes shown to the left).

Louis Pasteur (1822–1895)

Louis Pasteur (pass-TUR), a French chemist, began studying microbes in 1864. He was working on an important business in France: the fermentation of wine and vinegar. He noticed that certain microbes could cause food and drink to spoil. Pasteur discovered that different microbes cause different kinds of spoiling, but heat can kill many of these microbes. Today, because of his work, milk is heated to 71°C for 15 seconds to kill the microbe that causes tuberculosis. Using heat to kill microbes is now known as pasteurization in Pasteur's honor. Look at Figure 8 on the next page. The word "pasteurized" on milk sold in stores tells you that the milk is safe to drink.

Figure 8: Pasteurized Milk
Pasteurization kills microbes that may be present in milk.

In 1865, Pasteur was asked to help the silk industry of France, which was having problems with silk production. Silk is produced from threads spun by a worm known as the silkworm. Pasteur observed a microbe that was infecting the silkworms and the leaves they ate. When he recommended that the worms and their food be destroyed, the silk industry was saved.

Pasteur knew that some diseases were infectious. He suggested that microbes, which he referred to as "germs," could cause infectious diseases and were easily spread by people. This idea is the basis of the **germ theory of disease.**

Robert Koch (1843–1910)

Slowly, the role of microbes in causing infectious diseases began to be accepted. But there was still more work to do. Which microbe caused which disease? In 1876, Robert Koch (KOKE), a German doctor, identified the microbe that caused anthrax (AN-thraks), an infectious disease that was killing cattle. He later went on to identify the microbes that caused tuberculosis and cholera. Amazingly, he did all of his work in the four-room apartment that he shared with his wife.

The substance inside this dish is known as agar. Many kinds of microbes can grow on agar, which provides food for the microbes.

Koch developed a way to prove that a specific microbe caused a particular disease. In the case of anthrax, he injected healthy mice with blood taken from farm animals that had died of anthrax. He injected another group of healthy mice with blood taken from healthy farm animals. All of the mice injected with the blood from the infected animals died of anthrax. None of the other group of mice developed anthrax. He then showed that he could isolate anthrax microbes only from the mice that were injected with blood from infected animals. He did not find anthrax microbes in the healthy mice. In this way, Koch was able to provide scientific evidence that the anthrax microbe caused anthrax. Figure 9 summarizes his experiment.

Koch also created new ways to grow cultures of uncontaminated microbes. In particular, he developed agar (AH-gur), a gelatin-like substance which is used to grow microbe cultures. Agar is still used today, as you will find out in Activity 47, "Reducing Risk."

Figure 9: Koch's Experiment

Healthy mice

Inject with blood from cow with anthrax

Mice die: Anthrax microbes can be isolated from blood of infected mice

Healthy mice

Inject with healthy cow blood

Mice live: No anthrax microbes in blood of healthy mice

Florence Nightingale

Florence Nightingale (1820–1910)
Joseph Lister (1827–1912)
William Stewart Halsted (1852–1922)

Ideas such as those of Pasteur and Koch were very important in the field of medicine. Florence Nightingale, an English nurse, published her ideas on disease in 1860. At the time, the idea that cleanliness was important in preventing disease was not a common one. She was one of the first to recognize the value of cleanliness and recommended it as a part of good nursing. Her efforts improved sanitary practices in military hospitals and led to fewer soldiers dying from infections due to contaminated battle injuries.

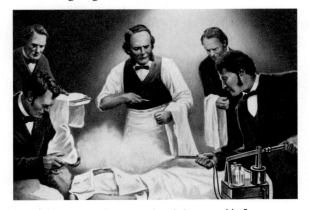

Joseph Lister supervises as antiseptic is sprayed before surgery

Scottish surgeon Joseph Lister had been concerned at the high death rates of patients following surgery. Surgery would be completed successfully, but about 45% of patients would die of infections afterward. When Lister heard about Pasteur's germ theory of disease, he came up with the idea of killing germs with chemicals. In 1867, he began using an antiseptic to clean surgical instruments. He also sprayed the air, and required hand washing and clean aprons. As a result, the death rate of patients following surgery dropped to 15%.

American surgeon William Halsted took these ideas one step further. Instead of just trying to kill the microbes once they were there, why not try to prevent them from being spread in the first place? In 1890, Halsted became one of the first surgeons to use rubber gloves during surgery. The gloves could be sterilized with heat and chemicals that were too hard on human hands. This helped reduce the presence of even more microbes and improve patient health.

By 1931, the germ theory of disease had become so accepted that ads for a disposable tissue read: "A new era in handkerchief hygiene! Use once and discard—avoiding self-infection from germ-filled handkerchiefs."

THE THEORY OF SPONTANEOUS GENERATION

Virchow stated that cells reproduce to create new cells. However, many scientists did not accept Virchow's ideas. They believed in spontaneous generation—the idea that living things grow from non-living things. For example, someone who believed in spontaneous generation might think that plants grow from soil. If you wanted to grow a plant, you would need only soil (no seeds or plant cuttings). After some time, a plant would spontaneously grow out of the soil. It took the experiments of many people to disprove the idea of spontaneous generation.

In 1668, an Italian doctor named Francesco Redi set out to show that maggots grew from eggs laid by flies. Because maggots were often found in rotting meat, many people believed that they just appeared spontaneously. To test his hypothesis, he set up several flasks containing meat. Some of the flasks were open to the air, some were sealed completely, and some were covered

Maggots are now known to be a juvenile stage of flies.

with gauze. As he expected, maggots appeared only in the open flasks in which the flies could reach the meat and lay their eggs.

➢

(continued from previous page)

In 1767, Italian priest Lazzaro Spallanzani conducted experiments to disprove spontaneous generation. He tightly sealed some bottles that contained liquid and then boiled them for more than 30 minutes. Nothing grew in the bottles. But because he had removed the air from the bottles using a vacuum, many scientists believed that Spallanzani proved only that spontaneous generation did not occur without air.

It was not until 100 years later, in 1859, that French chemist Louis Pasteur conducted a now-famous experiment that convinced most people. The French Academy of Sciences had sponsored a contest for the best experiment to either prove or disprove spontaneous generation. Pasteur's winning experiment was a variation of the method used by Spallanzani. He put a mixture of yeast, sugar, and water in several glass flasks. He then heated the necks of the flasks to bend them into the shape of an "S" (see Figure 10). Air could enter the flasks, but airborne microbes could not. Because of gravity, they would land somewhere along the neck of the flasks. Finally, he boiled the flasks to kill any microbes that might already exist in the mixtures. As Pasteur had expected, no microbes grew in the flasks. When Pasteur broke the neck of a flask and exposed it directly to air, microbes grew. Pasteur provided convincing evidence against the idea that living organisms come from non-living things.

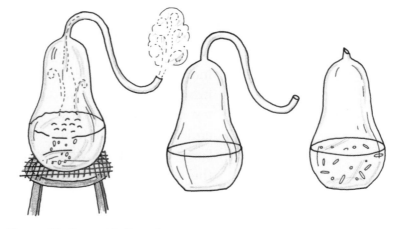

Figure 10: Pasteur's Experiment

ANALYSIS

1. Why is the germ theory of disease important in understanding infectious diseases?

2. How important was the development of the microscope in discovering the cause of infectious diseases?

3. **Reflection:** Imagine that each of the scientists in this activity wanted to hire an assistant. With which scientist would you most like to work? Why?

EXTENSION

Robert Hooke was an amazing scientist. His scientific ideas in the areas of physics, paleontology, biology, and chemistry are still relevant today. Why don't we know more about Hooke today? Some people believe it could be because the influential Sir Isaac Newton was his enemy. Find out more about Robert Hooke and his contributions to science. Begin by checking out links on the SALI page of the SEPUP website.

Schleiden, Schwann, and Siebold observed cells in plants, animals, and microbes. Since then, scientists have observed cells in every living organism. What do these cells look like?

CHALLENGE

What structures do different cells have in common? What structures are found only in some cells?

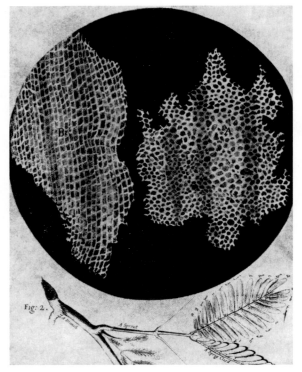

Robert Hooke's drawing of vegetable cells, 1665.

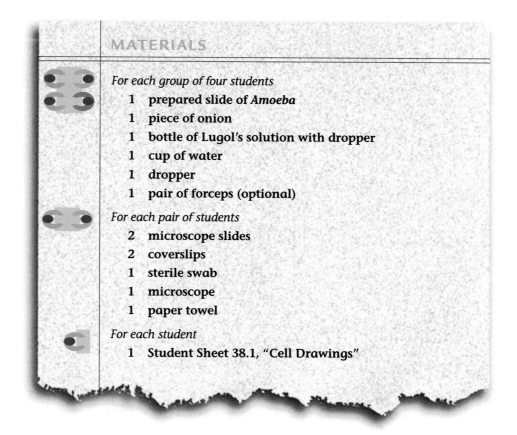

MATERIALS

For each group of four students

   1   prepared slide of *Amoeba*
   1   piece of onion
   1   bottle of Lugol's solution with dropper
   1   cup of water
   1   dropper
   1   pair of forceps (optional)

For each pair of students

   2   microscope slides
   2   coverslips
   1   sterile swab
   1   microscope
   1   paper towel

For each student

   1   Student Sheet 38.1, "Cell Drawings"

PROCEDURE

Within each group of four students, have one pair complete Parts One, Two, and Three in order. Have the other pair first complete Parts Two and Three, and then Part One.

Part One: Onion Cells

1. Use forceps or your fingernail to peel off a piece of the very thin inner layer of the onion.

2. Place 1–2 drops of water on a clean slide, then place your piece of onion in the drop of water.

3. Carefully place a coverslip over the cells. Begin by holding the coverslip at an angle over the water droplet, and then gradually lower the coverslip.

4. Be sure that your microscope is set on the lowest power (the shortest objective) before placing your slide onto the microscope stage. Center the slide so that the specimen is directly over the light opening and adjust the microscope settings as necessary.

Hint: To check that you are focused on the material that is on the slide, move the slide slightly while you look through the eyepiece—the material that you are focused on should move at the same time as you move the slide.

5. Observe the material on the slide.

Hint: If material on the slide is too bright to see, reduce the amount of light on the slide: do this by slightly closing the diaphragm under the stage. Move the slide so one or several of these cells are near the center of your field of view.

6. Without moving the slide (which can be secured with stage clips), switch to medium power (usually 10x). Adjust the microscope settings as necessary. Slowly focus up and down with the fine focus knob.

7. Without moving the slide, switch to high power (usually 40x). *Be careful not to smash the objective against the slide!* Adjust the microscope settings as necessary. Slowly focus up and down with the fine focus knob. You will see several layers of onion cells.

Hint: If material on the slide is too dark to see, increase the amount of light on the slide: do this by slightly opening the diaphragm under the stage.

8. Review "Microscope Drawing Made Easy" on page C-28 in Activity 36, "Looking for Signs of Micro-Life." Select one cell to draw at high magnification. Use Student Sheet 38.1, "Cell Drawings," for your drawings. Record the level of magnification next to your drawing.

9. Return to low power and remove the slide from the microscope. Add one drop of Lugol's solution onto the slide at the edge of the coverslip. Place a small piece of paper towel on the opposite side of the coverslip (see Figure 1); this will draw the stain under the coverslip.

10. Observe the cells again at low, medium, and high power and add any new details to your drawing. Record whether you can find the edge of the cell. Record your observations of the inside of the cell.

Figure 1: Staining the Slide

11. When you finish your observations, rinse the slide and coverslip and pat them dry with a paper towel.

Part Two: Cheek Cells

SAFETY

When you prepare your slide of cheek cells, each swab should be used by only one student. After spreading your cheek cells onto the slide, immediately discard your swab in the trash. Do not use it again. If you need to make another slide, use another swab.

12. Place 3 drops of water on a microscope slide.

13. Decide which person in your team of two students will volunteer to donate some cells from the inside of his or her cheek.

14. The volunteer should *gently* rub along the inside of his or her own cheek with a sterile swab. Turn the swab as you rub, making sure that each side of the swab is rubbed against the inside of your cheek. A very gentle scraping is sufficient—be sure not to cut or scratch your mouth!

15. Transfer the cheek scrapings onto the microscope slide by stirring the side of the swab in the water.

16. *Help prevent microbes from spreading!* Discard your swab in the trash as soon as you are done.

17. Carefully touch one edge of the coverslip to the water at an angle. Slowly allow the coverslip to drop into place.

18. Use Steps 4–8 as a guide to viewing and drawing your cheek cells.

19. Return to low power and remove the slide from the microscope. Add one drop of Lugol's solution onto the slide at the edge of the coverslip. Place a small piece of paper towel on the opposite side of the coverslip (see Figure 1 on the previous page); this will draw the stain under the coverslip.

20. Observe the cells again at low, medium, and high power and add any new details to your drawing. Record whether you can find the edge of the cell. Record your observations of the inside of the cell.

21. Remove the slide and follow your teacher's directions about where to put the slide for disinfecting.

Part Three: Microbe Cells

22. You and your partner should receive a microscope slide of *Amoeba*.

23. Use Steps 4–8 as a guide to viewing and drawing one *Amoeba* cell.

24. When you have completed all parts of the activity, turn off the microscope light and set the microscope back to low power.

ANALYSIS

1. Compare the three kinds of cells you have just observed.

    a. What structures do they have in common? Explain.

    b. How are the cells different? Explain.

2. Did you find evidence in this activity that the human body is made up of cells? Explain.

3. You stained the cheek and onion cells. How did the cells look before and after staining? Explain the purpose of the stain.

4. Do you think there are any small structures (organelles) inside your cheek cell other than the nucleus? What evidence do you have to support your answer?

LABORATORY

Now you know that all living organisms are made up of cells. Some are made of only a single cell. Others, such as people, onions, and elephants, are made of many cells. What do all these cells do?

Large multicellular organisms such as people take in oxygen. You use the oxygen to break down nutrients. This breakdown happens in the cells in organs all over your body. When your cells break down nutrients, wastes such as carbon dioxide are produced. In the picture on the left, the swimmer's lungs have taken in oxygen and are exhaling carbon dioxide. This oxygen is used to break down sugar from food in a process called **cellular respiration**. This process provides energy your body needs and releases carbon dioxide as waste.

How do we know that all these things happen in cells? In this activity, you will investigate yeast, another type of microorganism. Yeast is a single-celled organism. Using bromthymol blue (BTB) as the indicator, you will look for evidence that yeast cells respire. You may have used BTB in Activity 17, "Gas Exchange," when you investigated your own breath. Recall that BTB can be either blue or yellow. When there is carbon dioxide in a solution, BTB is yellow. Carbon dioxide is the main waste product of cellular respiration.

Like many other living organisms, people take in oxygen from the air and produce carbon dioxide.

CHALLENGE

What do yeast cells have in common with human cells?

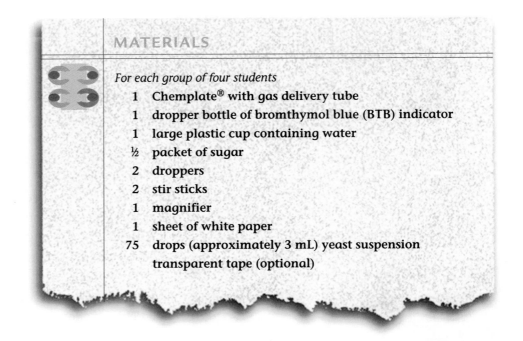

MATERIALS

For each group of four students

1	Chemplate® with gas delivery tube
1	dropper bottle of bromthymol blue (BTB) indicator
1	large plastic cup containing water
½	packet of sugar
2	droppers
2	stir sticks
1	magnifier
1	sheet of white paper
75	drops (approximately 3 mL) yeast suspension
	transparent tape (optional)

PROCEDURE

1. Have each person in your group of four choose one of the steps below and follow the instructions:

   a. Add 25 drops of yeast suspension to Cups 2, 3, and 4 of the Chemplate. Place this dropper near the yeast suspension. Do not use it for anything else.

   b. Add 2 drops of BTB indicator to the large oval cup of the Chemplate.

   c. Add 50 drops of water to the large oval cup of the Chemplate and stir with the BTB.

   d. Open the packet of sugar and carefully fill Cup 12 of the Chemplate with sugar. Give the rest of the sugar to another group of students.

2. Place the Chemplate on a sheet of white paper. Then use the small scoop on the end of the stir stick to add 10 scoops of sugar to the yeast suspension in Cup 3 and Cup 4.

3. Stir the sugar and yeast mixtures in Cup 3 and Cup 4.

4. Carefully observe the mixture in Cup 4 before capping it with the cup cover. Be sure that the cover fits tightly. (If not, use tape to secure it in place.)

5. Insert the tube extending off of the cup cover into the BTB solution (see Figure 1). If necessary, tape it into place.

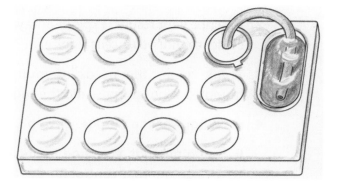

Figure 1: Placement of the Cover and Tube

6. *Create a standard for comparison:* Place 1 drop of BTB and 25 drops of water in Cup 8.

7. Create a data table to record your initial and final observations of Cup 2, Cup 3, Cup 4, Cup 8, and the large oval cup. Be sure to list what is in each cup—for example, Cup 2 (yeast + water). Then record your initial observations of each cup, including the initial color of the solution in the large oval cup.

8. Based on your knowledge of BTB as an indicator, predict what will happen in the large oval cup. Record your prediction, making sure to explain why you think this will happen.

9. Follow your teacher's directions for observing the yeast cells through a microscope.

10. After 10–15 minutes, make observations about the liquid in the large oval cup. Then stir the solution and record its final color, making sure to compare it to the color of the standard in Cup 8.

11. Observe the mixtures in Cups 2, 3, and 4. Notice whether (and how) they have changed. Enter your observations in your data table.

ANALYSIS

1. Compare your experimental results to your prediction. Was your prediction correct? Explain.

2. Describe your results. Explain how your results do or do not provide evidence that yeast cells respire.

3. Think about the needs of multicellular organisms such as humans. What purpose did the sugar serve for the yeast?

4. **a.** What was the purpose of Cup 2?

   b. Imagine that you had more materials available to you. Design another control for this experiment.

5. Based on your observations of the yeast cells under the microscope, your investigation of the gas produced by the yeast cells, and the picture of yeast cells at high magnification in Figure 2, what do yeasts have in common with humans?

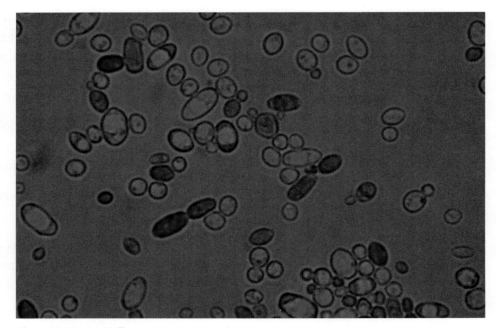

Figure 2: Yeast Cells

People, plants, and microbes—what do they have in common? They are all made of cells. By 1846, scientists realized that cells were not hollow shapes, like balloons, but were more solid, like gelatin. You may have observed several kinds of cells. In the cheek cell and microbe cell, you could see that there was material inside the cell that can be stained. The material that fills much of the inside of cells is called **cytoplasm** (SIGH-toh-pla-zum). Every cell also has a **cell membrane** that separates it from other cells and from the environment. You can see the cytoplasm and cell membranes of the stained skin cells shown in the photo below. You were probably able to see the cell membrane of your stained cheek cells.

How does a cell membrane work? Find out by creating a simple cell model. You will use a plastic bag to model the cell membrane.

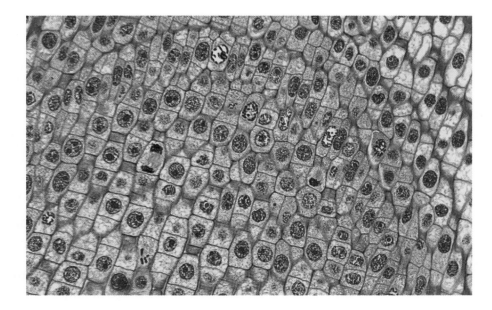

CHALLENGE

What is the function of a cell membrane?

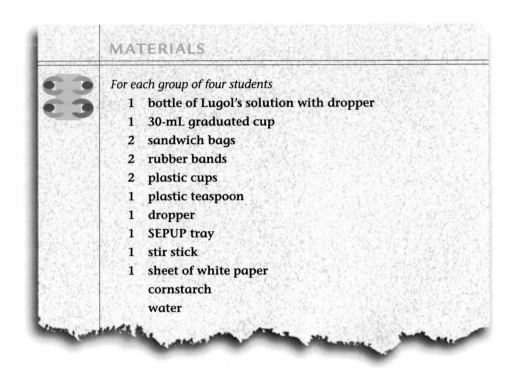

MATERIALS

For each group of four students

1 bottle of Lugol's solution with dropper
1 30-mL graduated cup
2 sandwich bags
2 rubber bands
2 plastic cups
1 plastic teaspoon
1 dropper
1 SEPUP tray
1 stir stick
1 sheet of white paper
   cornstarch
   water

PROCEDURE

1. Label the plastic cups as "Cup 1" and "Cup 2."

2. Use the graduated cup to pour 100 mL of water into each of these cups.

3. Add 7 drops of Lugol's solution to Cup 1.

4. Add 1 level teaspoon of cornstarch to Cup 2 and stir until mixed.

5. In your group of four, have one pair of students complete Step 5a, while the other pair completes Step 5b:

   a. Use the graduated cup to pour 30 mL of water into a sandwich bag. Then add 7 drops of Lugol's solution to the water in the bag.

   b. Use the graduated cup to mix 30 mL of water with one teaspoon of cornstarch. Stir and then carefully pour the mixture into a sandwich bag. Be careful to avoid getting cornstarch on the outside of the bag. If there is cornstarch on the outside of the bag, rinse the bag under cold water.

6. Use rubber bands to seal the bags.

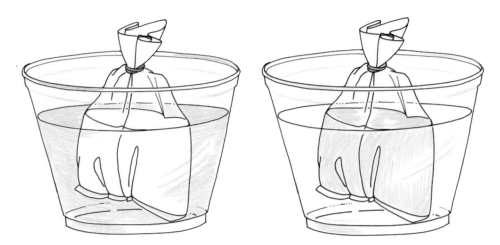

Figure 1: Initial colors of the mixtures.

7. Place the bag containing cornstarch into Cup 1 and the bag containing Lugol's solution into Cup 2, as shown in Figure 1. Then place the cups on the sheet of white paper and leave them there for 10–15 minutes. As you wait, complete Steps 8–10.

8. Create a data table to record the initial and final color of each solution, both inside and outside of the model cells in both cups. Be sure to record your initial observations.

9. Complete Steps 9a–c to find out how Lugol's solution reacts with starch.

    a. Place 5 drops of water into Cup 1 and 5 drops of water into Cup 2 of the SEPUP tray.

    b. Use the stir stick to add 1 scoop of cornstarch into Cup 2 and stir.

    c. Add 1 drop of Lugol's solution to each cup.

    d. In your science notebook, record the color of Lugol's solution when starch is present.

10. Complete Analysis Question 1.

11. After 10–15 minutes (or longer), lift the bags out of the cups and look carefully at all of the solutions. Record any changes that have occurred either in the bags or in the cups.

EXTENSION

Model a cell by using a real membrane from an egg. An egg can be "de-shelled" by soaking it in vinegar, leaving the rest of the egg intact. Be careful, the "de-shelled" egg is fragile. You can then place the de-shelled egg in different liquids, such as water, food coloring, paint, or corn syrup. Leave the egg in a solution for several days to find out if particles pass through the membrane. Collect data on these changes by measuring the mass of the egg before and after its soak.

ANALYSIS

1. **a.** Draw a diagram of the cell model used in this activity.

    b. Label the part of the cell model that represents the cell membrane and the part that represents cytoplasm.

    c. Label the part of the model that represents the environment outside the cell.

2. Review your results. Describe which part(s) of the lab set-up showed a reaction between Lugol's solution and starch.

3. Summarize your results by answering the following questions:

    a. Which particles—starch or Lugol's—were able to cross the model cell membrane? Explain how the experimental evidence supports your answer.

    b. Which particles—starch or Lugol's—were *unable* to cross the model cell membrane? Explain how the experimental evidence supports your answer.

4. Based on your cell model, what is the function of the cell membrane?

5. Think about the fact that cells are alive. Why is it important for particles to be able to pass through the cell membrane?

MODELING

Some organisms, like bacteria, consist of only one cell. Other organisms consist of several to many cells. An adult human being is made up of approximately 10 trillion cells. One drop of human blood, has about 500 *million* cells!

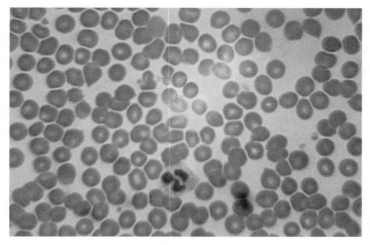

Red blood cells are the most numerous cells in blood.

Why do some cells need to be so small? Why aren't multicellular organisms like people made up of just one huge cell instead? Find out by modeling large and small cells.

CHALLENGE ➤

Why are cells so small?

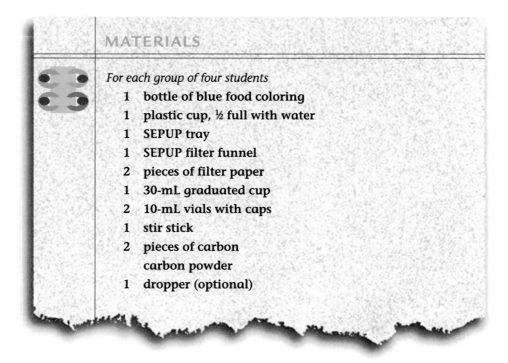

MATERIALS

For each group of four students

1	bottle of blue food coloring
1	plastic cup, ½ full with water
1	SEPUP tray
1	SEPUP filter funnel
2	pieces of filter paper
1	30-mL graduated cup
2	10-mL vials with caps
1	stir stick
2	pieces of carbon
	carbon powder
1	dropper (optional)

➤

REMINDER

Be careful when handling carbon. It is messy and can ruin your clothes. Never place any carbon directly onto a counter; use a piece of paper or a paper towel. Be sure to carefully clean up any spills.

PROCEDURE

1. Fold 2 pieces of filter paper into cones as shown in Figure 1: first fold each paper in half and then in half again. Open each filter paper into a cone (pull one piece to one side and push the rest to the other side).

Figure 1: Folding Filter Paper Into a Cone

2. Place the plastic SEPUP filter funnels over large Cups C and D of your SEPUP tray, as shown in Figure 2. Then place a filter paper cone into each of the funnels.

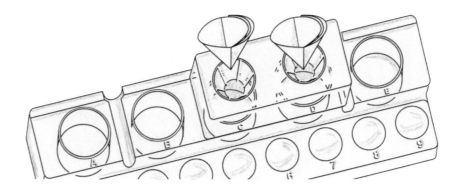

Figure 2: Setting Up the Filter

3. Dye your cup of water blue by adding 2 drops of blue food coloring. Stir.

4. *Model large cells:* Place 2 pieces of carbon into one of the 10-mL vials.

5. *Model small cells:* Using the scoop on a stir stick, your teacher will measure out the same volume of carbon powder into the other vial. You should now have the same amount of carbon in each of the two vials.

6. *Model how well the cells can take up oxygen or nutrients they need to live:* Use your 30-mL cup to add 7.5 mL of dyed water to each vial. Then cap the vials and shake each vial ten times.

7. Open the vial containing the carbon pieces. Pour the mixture through the filter paper over Cup C.

8. Open the vial containing the carbon powder. Pour the mixture through the filter paper over Cup D.

9. Observe and record the color of the water in each large cup of your SEPUP tray.

10. Clean up as directed by your teacher.

ANALYSIS

1. In this model, what did each of the following represent:

    a. carbon powder

    b. carbon pieces

    c. blue dye

2. What happened to the blue dye in each vial? Explain.

3. According to the model, which cells—large or small—are most efficient at taking up oxygen and nutrients from the environment? Explain.

4. What is one reason multicellular organisms, such as people, are made up of many small cells instead of one large cell?

You learned that Schleiden and Schwann discovered that all living organisms are made up of one or more cells. This includes plants, animals, and many microbes. The microbes that cause infectious disease are often organisms made up of just one cell, as Siebold discovered. Most organisms you can see without a microscope are made of many cells.

The cells shown here are from human skin. What do you see inside the cells? What exactly are you looking at when you use a microscope to look at a cell?

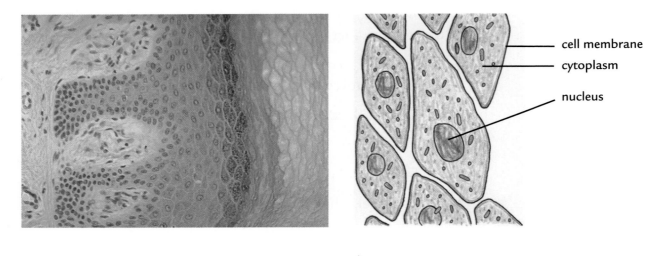

cell membrane
cytoplasm
nucleus

CHALLENGE

What are some of the parts of a cell? What do they do?

READING

A Typical Cell

As scientists continued to study cells, they noticed structures inside the cells. From observing many kinds of cells in thousands of organisms, biologists discovered that some structures are found in all or nearly all cells. These structures are so common that they are usually included in models or drawings of a "typical cell."

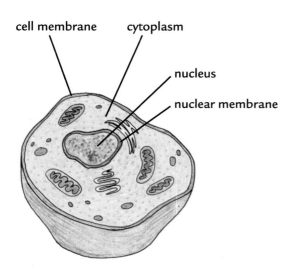

cell membrane   cytoplasm

nucleus

nuclear membrane

Figure 1: Inside a Cell

The most common structure of cells is the cell membrane (see Figure 1). This membrane separates the cell from its environment. Every kind of cell, whether a bacterium, or a cell from an elephant or from a giant oak tree, has a cell membrane. As you learned in Activity 40, "A Cell Model," the cell membrane acts as a barrier to control what enters or leaves the cell. Somehow, everything that enters or leaves the cell must cross this membrane.

The material enclosed by the cell membrane is called the cytoplasm, which means the "cell material." In the cytoplasm the cell breaks down nutrients from food and builds the new substances it needs to grow and to carry out its other functions.

STOPPING TO THINK 1

a. How did scientists discover the common structure of cells?

b. What are some of the common structures of a cell?

In 1831, Robert Brown identified a small dark center within many cells. He called this center the **nucleus.** You were probably able to observe the nucleus in onion, *Amoeba,* and human cells. Most organisms—except for bacteria—have a cell nucleus. The nucleus is a small compartment within the cell. It is separated from the rest of the cell by a nuclear membrane. The nucleus contains the genetic information of the cell and directs the cell's activities, including growth and reproduction.

STOPPING TO THINK 2

a. Why is the nucleus an important part of most cells?

b. What type of organism does not contain a nucleus?

Most cells have other tiny structures that help them do many jobs. These structures are called **organelles,** or "little organs." They are often surrounded by their own special membranes. Some of the organelles can just barely be seen with a light microscope. Some of the jobs performed

➢

by these organelles include obtaining and storing energy, helping cells move and divide, and making substances that are either used in the cell or transported to other parts of the body.

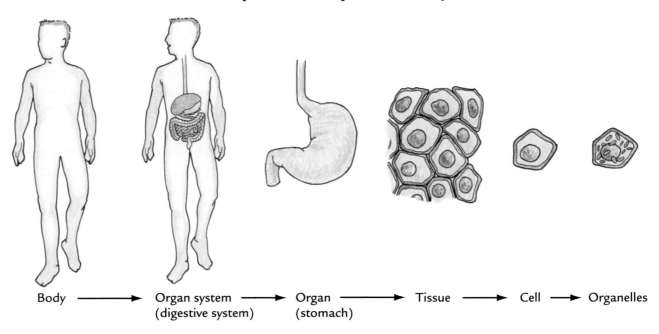

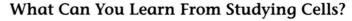

Body ⟶ Organ system ⟶ Organ ⟶ Tissue ⟶ Cell ⟶ Organelles
(digestive system)   (stomach)

What Can You Learn From Studying Cells?

Information about cells can be used for practical purposes, such as treating different kinds of diseases. Here are just two examples:

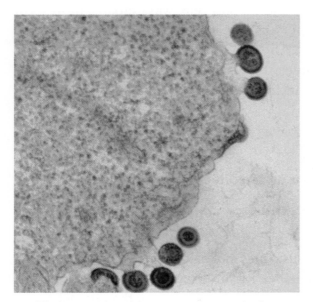

A cell biologist used an electron microscope to take this photo of HIV (the virus that causes AIDS) on the surface of a human white blood cell.

- AIDS is a disease of one group of cells within the human body. Investigating how these cells work normally and what goes wrong in the cells of a person with AIDS helps scientists understand the disease and develop treatments.

- Human blood contains red blood cells. Each red blood cell lives in the human bloodstream for only about 120 days. New red blood cells are constantly being formed as old ones die. Researchers have used information about how new red blood cells develop to prepare a drug that causes more red blood cells to form. This drug is given to patients who require certain kinds of surgery or to people with illnesses that reduce the number of red cells in the blood. The drug helps patients build up red cells and require fewer blood transfusions.

CELL BIOLOGY

Cell biology is the special branch of biology that studies cells and how they work. Cell biologists are fascinated by the variety of cells and the fantastic structures inside them. Some cell biologists study one type of cell, such as muscle cells. Others focus on special parts of cells, such as cell membranes, to understand how they work. Some of the questions that cell biologists try to answer are:

- How do the different cells in an organism work?

- What does each part of the cell do, and how?

- How do the different cells in an organism communicate and control their activities so that things happen in the right place and at the right time?

- How can one fertilized human egg cell grow into a complex adult with many kinds of cells?

- How does a cell know when to divide?

Scientists have found some exciting partial answers to these questions, but there is a great deal left to learn about cells.

A cell biologist using an electron microscope

For links to photographs and more information about cells, go to the SALI page of the SEPUP website.

ANALYSIS

1. Observe the pictures of cells in Figure 2, "Animal Cells." Cells 1, 2, and 4 were taken with a scanning electron microscope which shows the surface (and not the inside) of the cell. This type of microscope magnifies the cells much more than the microscopes you use in class. You can see that the cells have quite different

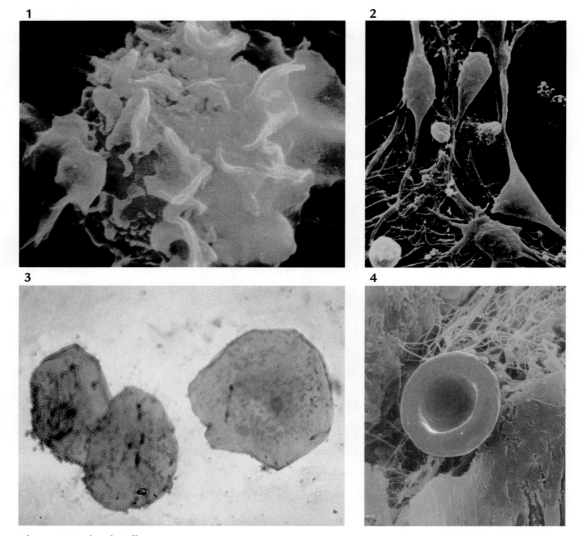

Figure 2: Animal Cells

shapes: some are rounded, while others are elongated, flat, or ruffled. These shapes depend on the cells' functions in the body. Try to match each cell with one of the following descriptions.

   a. These cells have long branching parts that send signals to distant parts of the body.

   b. These flat cells form an even covering on the surface of areas like the inside of the mouth.

   c. These round human cells are unusual because they do not have a nucleus. They are full of a protein that carries oxygen to all parts of the body.

   d. These cells are able to crawl around the body to attack bacteria and other foreign material. Ruffles on the cell membrane lead the way as the cells move.

2. Based on its description, which of the four cells described in Question 1 is a nerve cell? Which is a red blood cell? Which is a white blood cell? Which is a skin cell? Explain how you were able to match the type of cell with its function.

3. Give one example of how the study of cells helps treat diseases.

4. Explain why membranes are so important to cells.

5. Look back at your drawings from Activity 36, "Looking for Signs of Micro-Life." Did you observe any structures within the microbes that you drew? What do you think these structures are?

6. **Reflection:** Which of the questions studied by cell biologists is most interesting to you? Why?

Study the photographs on this page. You can see microbes of different shapes, sizes, and structures. Microbes are organized into different groups based partly on differences in their cell structure. In this activity, you will look at two different groups of microbes to see what kinds of differences you can find. You will observe stained slides of **protists** (PRO-tists) and **bacteria** (bak-TEER-ee-uh).

You first saw microbes when you looked at water samples in Activity 36, "Looking for Signs of Micro-Life." Do you recognize any of the same creatures in these photographs?

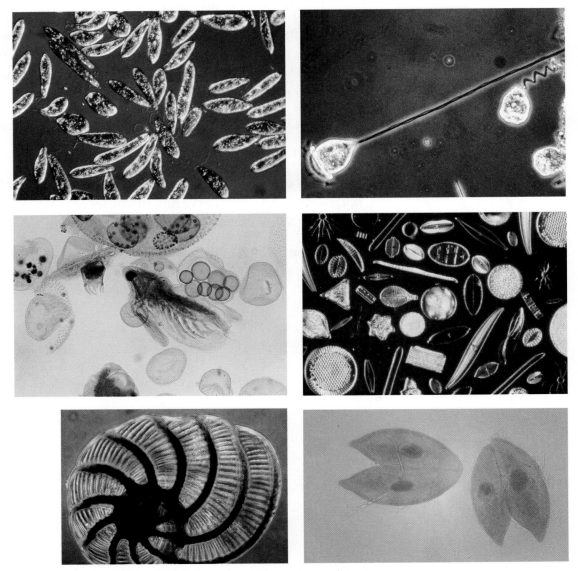

Microbes display a fascinating variety of shapes, sizes, and structures.

CHALLENGE

What are some of the differences among the cells of two groups of microbes?

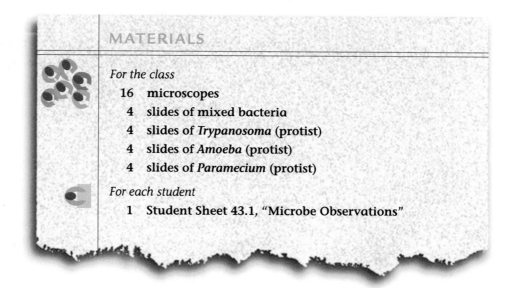

MATERIALS

For the class
16 microscopes
4 slides of mixed bacteria
4 slides of *Trypanosoma* (protist)
4 slides of *Amoeba* (protist)
4 slides of *Paramecium* (protist)

For each student
1 Student Sheet 43.1, "Microbe Observations"

PROCEDURE

1. You and your partner should receive a microscope slide of a one-celled microbe.

2. Be sure that your microscope is set on the lowest power (shortest objective, usually 4x) before placing your slide onto the microscope stage. Center the slide so that the specimen is directly over the light.

3. Begin by observing the slide on low power. You may need to search the slide for the organisms. Be sure that an organism is in the center of the field of view (you may need to move the slide slightly) and completely in focus before going on to Step 4.

   Hint: To check that you are focused on the material that is on the slide, move the slide slightly while you look through the eyepiece—the material that you are focused on should move at the same time as you move the slide.

   Hint: On prepared slides, organisms are usually stained with dyes to make them easier to see: look for blue, purple, green, or pink organisms.

4. Without moving the slide (which can be secured with stage clips), switch to medium power (usually 10x). Adjust the microscope settings as necessary. Observe the organism.

5. Without moving the slide, switch to high power (usually 40x). *Be careful not to smash the objective against the slide!* Adjust the microscope settings as necessary.

   Hint: If material on the slide is too dark to see, increase the amount of light on the slide: do this by slightly opening the diaphragm under the stage.

6. Turn the fine focus knob up and down just a little to reveal details of the microbe at different levels of the slide.

7. Review "Microscope Drawing Made Easy" on page C-28 of Activity 36, "Looking for Signs of Micro-Life." Then draw your organism (on high power) on Student Sheet 43.1, "Microbe Observations." Be sure to record the level of magnification you are using. Make your drawing large enough to fill up most of the space on the paper. Include on your drawing details inside the cell and along the edge of the cell membrane.

8. Switch slides with another pair of students and repeat Steps 2–7.

9. Repeat Step 8 until you have seen all four microbe slides.

10. When you have completed your observations, turn off the microscope light and set the microscope back to low power.

11. Work with your group to discuss Analysis Questions 1 and 2 before the class discussion.

ANALYSIS

1. When you compare the different protists, what differences do you observe?

2. When you compare the different bacteria, what differences do you observe?

3. When you compare all of the different microbes, what similarities and differences do you observe?

4. Look at the drawings of micro-life you made for Activity 36, "Looking for Signs of Micro-Life." Could any of the organisms you saw have been protists or bacteria? Support your answer with evidence from this activity.

5. In your science notebook, create a larger version of the diagram shown below (known as a Venn diagram). Record unique features of cells of each group of organisms in the appropriate space (either "protists," "bacteria," or "human" cells). Record common features between groups in the space that overlaps. **Hint:** Think about what you have learned about cells in the last few activities. Look again at your notes from this activity.

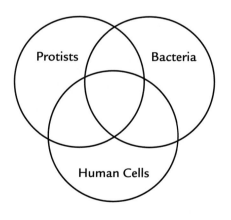

INVESTIGATION

As you will learn in the next few activities, diseases caused by different microbes are prevented and treated differently. That's why Leeuwenhoek's discovery of microbes and Pasteur's germ theory of disease were essential. Find out more by using more evidence to classify microbes.

CHALLENGE

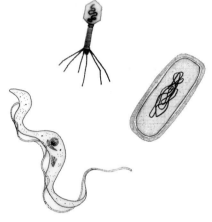

How are these microbes classified?

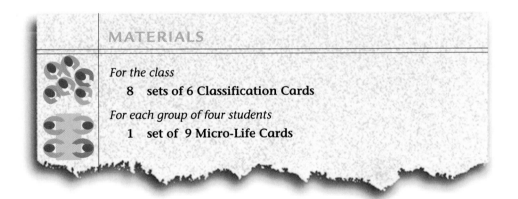

MATERIALS

For the class

   8   sets of 6 Classification Cards

For each group of four students

   1   set of 9 Micro-Life Cards

PROCEDURE

1. Spread your Micro-Life Cards out on a table. Each card shows a high magnification of the outside of the microbe and a drawing of a high magnification view of the inside.

2. Examine each card carefully, noting similarities and differences.

3. With your group members, classify the microbes into groups. Work together to agree on a classification system:

   • Listen to and consider the explanations and ideas of other members of your team.

   • If you disagree with other members of your team about how to classify a microbe, explain why you disagree.

4. In your science notebook, list the groups that you created and the common features of each group. Be sure to record which microbes belong to which group.

5. Leave your cards sorted into groups and your notebook open on your work surface. When all teams are finished you will look at what others have done.

6. View the work of other student teams. As you look at their classification systems, observe the similarities and differences between their systems and your own. Discuss your observations with your team members.

7. You will receive 6 Classification Cards from your teacher. Each card represents a group of creatures. Based on the information described on the Classification Cards, place each Micro-Life Card under one of the Classification Card categories. In your science notebook, record any changes you want to make to your original grouping of your Micro-Life Cards.

8. As a class, discuss the classification of the Micro-Life Cards. In your science notebook, record the common features of each major group.

ANALYSIS

1. How could knowing the structure and classification of disease-causing microbes help scientists fight a disease?

2. How did your system of classification compare to the Classification Cards?

3. Look back at the generalized animal cell in Figure 1 in Activity 42, "A Closer Look," on page C-59. Explain how this drawing of a cell is similar to or different from the structure of each of the following groups of microbes:

   a. protists

   b. bacteria

   c. viruses

Have you had a cold, flu, or other infectious disease recently? Do you know what caused your illness? Microbes cause most infectious diseases. Microbes include the protists, bacteria, and viruses that you classified in Activity 44, "Who's Who?" They also include some fungi, such as yeast and the fungi that cause athlete's foot.

By now you know that *germ* is simply another word for a microbe that causes disease. But you may have also heard the word *microorganism* used. Why, then, do we keep referring to microbes? To find out, you need to know a little more about the differences among the microbes you've studied so far (protists, bacteria, and viruses).

CHALLENGE

How do microbes fit into the classification of organisms?

READING

Classifying Organisms

Until recently, scientists classified organisms into five groups, called Kingdoms, as shown in Figure 1. New evidence has led to several alternatives to the five-kingdom system. Classification is a way to make sense of a lot of information. As the information changes, new classification systems evolve. For example, scientists have learned that bacteria can be divided into two very different groups, called Bacteria and Archaea. Still, it is useful for you to think about five different groups of organisms: animals, plants, fungi, protists, and bacteria.

Figure 1: The Five-Kingdom Classification Scheme

Animals	Plants	Fungi	Protists	Bacteria

You are most familiar with animals and plants. They make up two kingdoms. A third kingdom is made up of fungi. The fungi include yeasts (like the one you used in Activity 39, "Cells Alive!"), molds, and mushrooms. Protists and bacteria, like the ones you observed in Activity 43, "Microbes Under View," belong to two more groups of organisms. Notice that viruses are not included in the figure because they are not considered to be living organisms.

STOPPING TO THINK 1

Think about all of the slides you have observed. Have you observed cells of organisms from every kingdom? List all the cells you have observed from organisms in each kingdom.

Protists

Protists are single-celled microbes that have a nucleus. While some protists cause illness, many others are harmless. The *Trypanosoma* that you observed in Activity 43 is closely related to another type of *Trypanosoma* that causes sleeping sickness in people. Species of *Paramecium* are often harmless, living in fresh and salt water, where they feed on bacteria, algae, and other protists. Many types of *Amoeba* are harmless, while others cause illnesses of the digestive system.

Bacteria

Bacteria are single-celled microbes that do not have a nucleus. Bacteria are also the most common microbes and can be found everywhere—in snow, deserts, lakes, the ocean, and the human body. As you may recall, bacteria are extremely tiny; a thousand bacteria could fit in a cluster on the dot of an "i." There are more bacterial cells in your digestive system and on your skin than the number of cells that make up your entire body!

While some bacteria, such as *Mycobacterium tuberculosis*, cause diseases, other species of bacteria are helpful. In fact, without bacteria, nothing would ever decompose; the world would be full of dead organisms, from the tiniest microbes to large plants and animals! Bacteria also are important in the preparation of foods and beverages. You may have noticed a statement on some yogurt containers: "contains live and active yogurt cultures." That's because yogurt is produced by the fermentation of milk by bacteria! Figure 2 shows the shapes and some information about different kinds of bacteria.

STOPPING TO THINK 2

Would you describe bacteria as being helpful or harmful to people? Explain.

Figure 2: Some Common Types of Bacteria

Shape	Examples	Ecological Roles
sphere 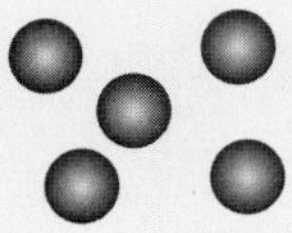	*Diplococci* (pairs of cocci) 	cause pneumonia
	Staphylococci (clusters of cocci) 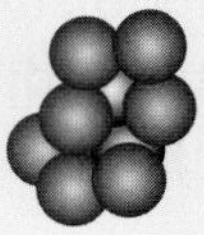	are normally present on human skin; some cause boils and infections
	Streptococci (chains of cocci) 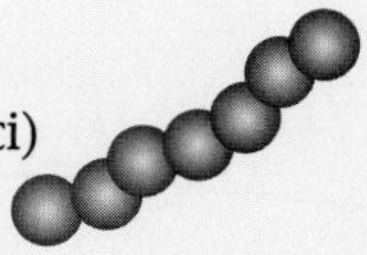	are used to make yogurt and cheese; cause strep throat
rod 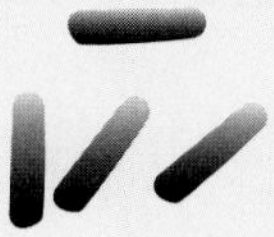	*Bacilli* (rods)	decompose hay; are used to make cheese, yogurt, pickles, and sauerkraut; are normally present in the human digestive tract; cause diarrhea; cause anthrax in cattle and sheep
	Mycobacteria (chains of bacilli) 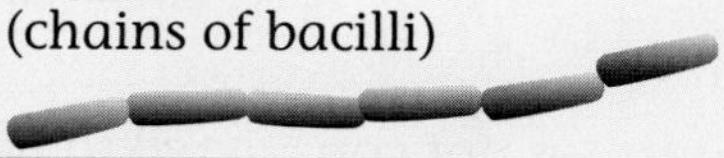	cause tuberculosis; are found normally in soil and water.
curved rod 	*Vibrio*	cause cholera; help break down sewage
short spirals 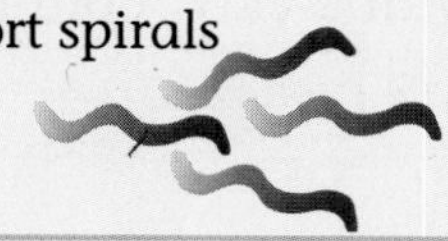	*Spirilla*	are decomposers in both fresh and salt water
long spirals 	*Spirochete*	cause syphilis; are decomposers
branched chain 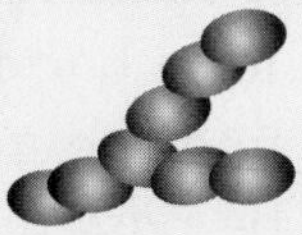	*Actinomyces*	produce several antibiotics; were once classified as fungi

Cocci are spherical bacteria; the singular of cocci *is* coccus.

Viruses: A Group Apart

Viruses are not living organisms. Unlike protists, bacteria, and all other living organisms, viruses are not made up of cells. They are unable to grow or reproduce independently or carry out the functions, such as respiration, that living organisms do. Instead, viruses rely on the cells of living organisms for their reproduction. It is for this reason that we say infectious diseases are caused by microbes, and not microorganisms.

Figure 3: Comparing Average Sizes of Microbes
These are relative, not actual, sizes of microbes. An average bacterium is actually much smaller than the virus shown here.

STOPPING TO THINK 3

a. Why are viruses not considered to be microorganisms?

b. Look at Figure 3, "Relative Sizes of Microbes." How do the sizes of protists, bacteria, and viruses compare?

c. Which do you think cannot be seen with a classroom microscope?

How do we know viruses exist? The existence of viruses was first suggested in 1898, nearly 45 years before they were first seen. In 1895, Dutch scientist Martinus Beijerinck (BY-er-ink) began experimenting with the tobacco plant. He was studying a plant disease that he believed to be infectious. By this time, scientists were familiar with protists and bacteria, so Beijerinck began searching for a bacterium that might be causing this disease. But he could not find one. Yet his experiments demonstrated that the disease could be passed from plant to plant, so he concluded that the disease was caused by a microbe. Since it wasn't a protist or a bacterium, he called it a virus, which means "poison" in Latin.

Viruses are so small that you need an electron microscope to see one. The electron microscope was not invented until the 1930s. As a result, viruses were first seen in 1939. Today, we know that viruses cause many diseases, including the flu, colds, chickenpox, and AIDS.

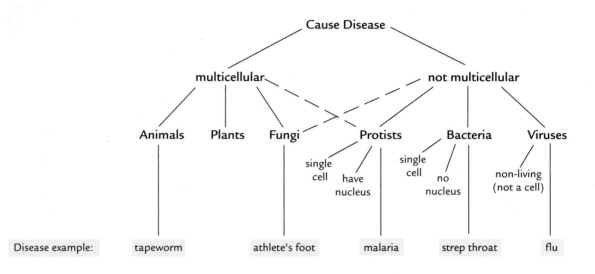

Figure 4: Classifying Disease-Causing Organisms and Viruses

Figure 4 shows the five-kingdom classification plus viruses. Note the examples of diseases caused by members of each group. What do you think the dotted lines mean?

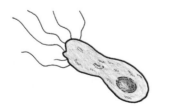

For links to more information about microbes, go the SALI page of the SEPUP website.

ANALYSIS

1. You have read how microbes can be both helpful and harmful to humans. Do you think a microbe can be *neither* helpful nor harmful? Explain.

2. You decide to examine some pond water under a microscope. With a magnification of 40 (using the 4x objective), you observe a long, cylindrical organism moving across your field of view (see left). As you look more closely, you notice what appears to be a round structure inside of it. Is this organism most likely a protist, bacterium, or virus? Explain how you arrived at your conclusion.

➢

3. Suppose your school's microscopes did not have 40x objectives, but only 10x objectives. Your friend, who is in high school, uses a 40x objective. Explain what group of microbes he or she can study that you cannot.

4. What are the advantages of using the highest power objective on a microscope? What are the advantages of using the lowest power objective on a microscope? Explain.

5. In your science notebook, draw a larger version of the Venn diagram shown below. Record unique features of each group of microbes in the appropriate space. Record common features among groups in the spaces that overlap. **Hint:** Think about what you have learned about cells in the last few activities.

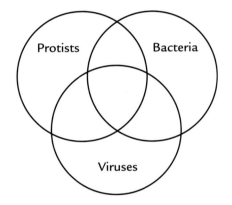

6. **Reflection:** On a field trip, you visit a laboratory that has an electron microscope. The microscopist (the person who runs the microscope) offers to set up a microbe for you to view. What microbe, or group of microbes, would you choose to view? Why?

What does your body do to protect itself from invading microbes? Even before an organism can enter your body, your skin provides a protective barrier. But foreign substances can still enter through cuts or natural body openings, such as your mouth or your nose. Tears, saliva, and mucus help to remove some invaders at these sites. But when foreign substances cross these barriers, your **immune** (ih-MYOON) **system** comes to the rescue.

Your immune system has the amazing ability to distinguish between the substances of your own body and foreign substances, such as bacteria and viruses. A healthy immune system can then mount a defense against these invaders. Several kinds of cells, particularly white blood cells, are responsible for this immune response. The pictures here show normal human blood cells. Note that the red blood cells are the most common. Also note the detail of the white blood cells. They increase in number when the body is under attack from a foreign substance.

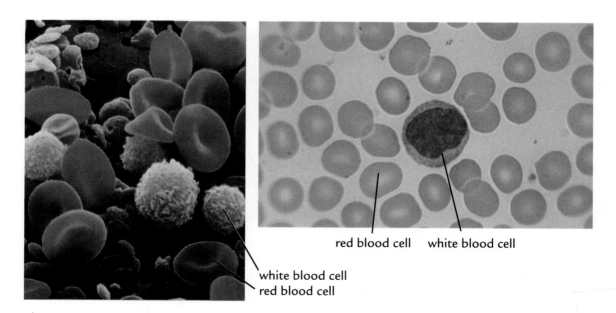

red blood cell   white blood cell

white blood cell
red blood cell

The photograph on the left was taken through a scanning electron microscope, while the photograph on the right was taken through a light microscope.

Immune responses of the human body are not always helpful. Any new material in the body, including blood and organs, can trigger an immune response. It is this reaction of the immune system that makes organ transplants and blood transfusions difficult. If the blood type of the blood donor is not compatible with that of the person receiving the blood, the transfused blood cells are seen as foreign by the immune system and they clump together. These clumps can create blockages in blood vessels and cause death. That is why it's important to know which types of blood can be donated safely to people with each of the four human **blood types: A, B, AB, and O.** You will simulate what happens to a person's blood when blood from a donor is added.

• • •

For links to more information on the blood and diseases of the blood, go the SALI page of the SEPUP website.

CHALLENGE

How does your blood help fight infectious diseases?

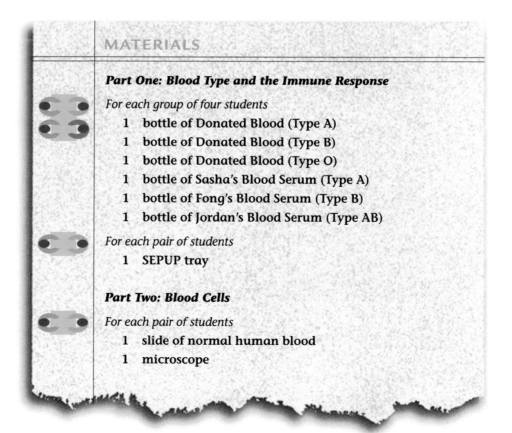

MATERIALS

Part One: Blood Type and the Immune Response

For each group of four students
1   bottle of Donated Blood (Type A)
1   bottle of Donated Blood (Type B)
1   bottle of Donated Blood (Type O)
1   bottle of Sasha's Blood Serum (Type A)
1   bottle of Fong's Blood Serum (Type B)
1   bottle of Jordan's Blood Serum (Type AB)

For each pair of students
1   SEPUP tray

Part Two: Blood Cells

For each pair of students
1   slide of normal human blood
1   microscope

PROCEDURE

Within each group of four students, one pair begins with Part One and the other pair begins with Part Two. When both pairs have completed their parts, they can switch roles.

Part One: Blood Type and the Immune Response

BLOOD EMERGENCY!

Three patients needing blood transfusions have arrived at the local hospital. This is the chart showing their blood types. In order to supply the blood, the hospital staff has asked the community to help. Several people respond by donating blood. The hospital receives blood donations of types A, B, and O, but these blood types might not be compatible with each patient.

Patient	Blood Type
Sasha	A
Fong	B
Jordan	AB

Does the hospital have enough of the right type of blood for each patient? Find out by testing samples of each blood type.

1. Collect the three blood samples and the three serum samples.

    Note: *Serum* is blood that has had the red blood cells removed. In blood transfusions, the donor's blood must be compatible with the patient's serum.

2. Design a data table to record your experimental results. You will test each of the three donated blood types with serum from each of the three patients.

3. Place two drops of Sasha's Blood Serum in Cups 1–3 of your SEPUP tray.

➤

4. Add two drops of Donated Blood Type A to Cup 1. Record the results in your data table.

5. Test Sasha's Blood Serum with the remaining donated blood samples. Record the results in your data table.

6. Use Cups 4–9 to test the samples from the other two patients, Fong and Jordan. Record the results in your data table.

Part Two: Blood Cells

7. You and your partner should receive a microscope slide of normal human blood.

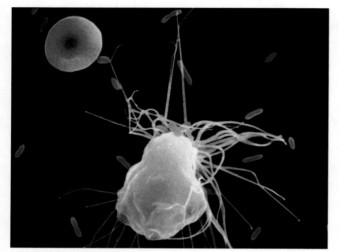

This high power scanning electron microscope photograph has been colorized. A red blood cell is near the top of the picture. A white blood cell (colored purple) is attacking bacteria (colored blue-green).

8. Be sure that your microscope is set on the lowest power (shortest objective) before placing your slide onto the microscope stage. Center the slide so that the specimen is directly over the light opening and adjust the microscope settings as necessary.

   Hint: To check that you are focused on the material that is on the slide, move the slide slightly while you look through the eyepiece—the material that you are focused on should move at the same time you move the slide.

9. Begin by observing the slide on low power (usually the 4x objective). Scan the slide and focus on a section that shows more than one kind of cell.

   Hint: Remember that stains are often used to make structures on a slide more visible. Look carefully for a light pink smear with a dark purple blob. If material on the slide is too light to see, reduce the amount of light on the slide: do this by slightly closing the diaphragm under the stage.

10. Without moving the slide (which can be secured with stage clips), switch to medium power (usually 10x). Adjust the microscope settings as necessary.

11. Without moving the slide, switch to high power (usually the 40x objective). *Be careful not to smash the objective against the slide!* Adjust the microscope settings as necessary.

Hint: If material on the slide is too dark to see, increase the amount of light on the slide: do this by slightly opening the diaphragm under the stage.

12. In your science notebook, describe the two different kinds of cells that you see. In your description, include which type of cell is more common, the shape of each cell, the relative size, and any cell structures you are able to identify in either cell.

ANALYSIS

Part One: Blood Type and the Immune Response

1. Each patient required one pint of blood. The hospital received one pint each of type A, B, and O blood. Explain whether the hospital had enough of the right type of blood for each patient.

2. What prevents your body from accepting transfusions of certain types of blood?

Part Two: Blood Cells

3. Think back to all the work that you have been doing on cells. Compare and contrast different types of cells by copying and completing the table below.

4. In what ways does your body prevent you from catching an infectious disease?

Cell Type	Cell Shape	Cell Membrane?	Cytoplasm?	Nucleus?
Bacteria				
Protist				
Plant (onion)				
Animal: cheek				
Animal: red blood cell				
Animal: white blood cell				

LABORATORY

How do you prevent yourself from catching an infectious disease? You now know that your immune system provides you with natural defenses, but sometimes your immune system becomes overwhelmed by disease-causing microbes and you get sick. One way to reduce your risk of getting sick is by taking simple precautions, like washing your hands before you eat. You may even use antimicrobial solutions, such as an antibacterial soap or a disinfectant, when you clean up. How effective are these products at killing germs?

You can measure the effect of different solutions on the growth of microbes. You can culture, or grow, microbes in a special dish known as a **petri** (PEE-tree) **dish**. The petri dish contains food for the organisms you are trying to culture. In classrooms, the most commonly used food is agar, a gelatin-like material that was first invented by Robert Koch. If bacteria are present, many of them will grow on the agar. You can see this in the petri dish in the photograph below. It is also possible for molds and algae to grow on agar.

CHALLENGE

How effective are different solutions at preventing the growth of microbes?

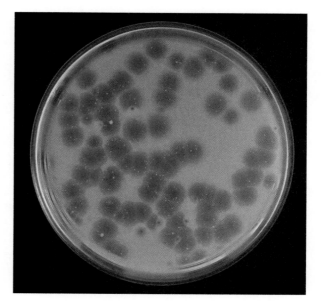

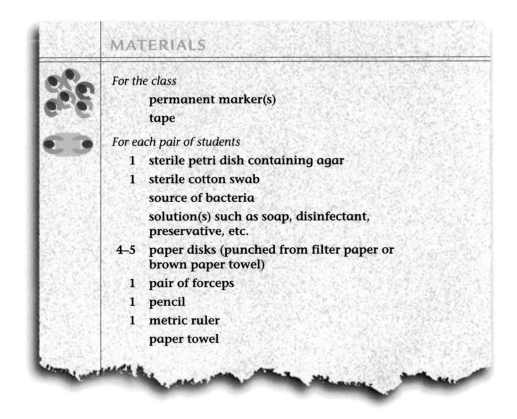

MATERIALS

For the class

    permanent marker(s)

    tape

For each pair of students

  1 **sterile petri dish containing agar**

  1 **sterile cotton swab**

    source of bacteria

    solution(s) such as soap, disinfectant, preservative, etc.

4–5 **paper disks (punched from filter paper or brown paper towel)**

  1 **pair of forceps**

  1 **pencil**

  1 **metric ruler**

    paper towel

PROCEDURE

1. Use a permanent marker to put your name and class period on the bottom of your petri dish.

2. Dip your swab in the source of bacteria provided by your teacher.

3. Remove the lid of the petri dish. Use the swab to streak the bacteria onto agar as shown in Figure 1. Press firmly but not too hard. Your goal is to spread the bacteria but not to break up the agar layer. Turn the dish 90° and repeat as shown in Figure 2.

Figure 1: First Streak

Figure 2: Second Streak

4. Use a pencil to write the initial of the solution you are testing on a disk of filter paper (or brown paper towel).

5. Use your forceps to dip the disk in the solution. To remove excess solution, touch the edge of the disk to a clean paper towel.

6. Place the disk on the agar of your petri dish and re-cover the dish.

7. Repeat Steps 3–6 for each solution you are testing. Be sure to leave at least 1.5 cm between the paper disks. Tape around the dish to seal it.

8. Place your petri dish in a warm place for a few days.

9. Check the growth of microbes in the dish each day. Record if the growth of microbes has stopped in the area around the disk. If so, use a ruler to measure this space, known as the zone of inhibition.

10. Examine the control dishes set up by your teacher. Record your observations.

ANALYSIS

1. Did the solution(s) affect the growth of bacteria on your petri dish? Explain. Be sure to compare your results to the control and to describe your evidence.

2. Share your results with the class.

   a. Did everyone who tested the same solution get the same results? Explain.

   b. How effective were the different solutions in preventing microbial growth?

   c. How might you follow up on this investigation or improve the design of this investigation?

3. **Reflection:** Have the results of your experiment caused you to want to change any of your behavior (such as what solutions you use to wash your hands or household surfaces)? Why or why not?

INVESTIGATION

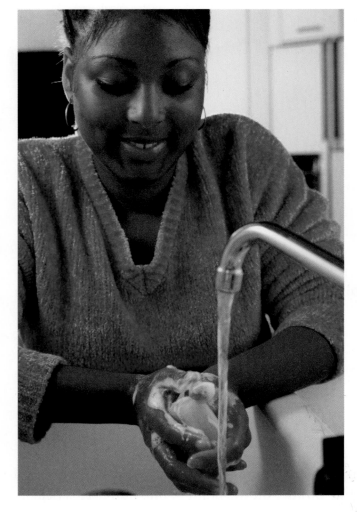

In the 1840s, Dr. Semmelweiss found that hand washing could significantly reduce the rate of infection in hospitals. One common type of illness that can be reduced by hand washing is food poisoning. Millions of people suffer from some form of food poisoning each year. Most people who get these infections don't die, but they feel terrible and miss work or school days.

Could hand washing reduce the number of times you get sick? Hands have about 200 million microbes on them. Most are harmless, but some of these microbes can cause food poisoning, colds, flu, and other infections. In fact, public health researchers estimate that 80% of common infections in the U.S. are caught by touching surfaces that are contaminated with infectious microbes. Contaminated surfaces might include sinks, countertops, doorknobs, or your own hands.

CHALLENGE ➡

How effectively does hand washing reduce the spread of microbes? How can you improve the effectiveness of hand washing?

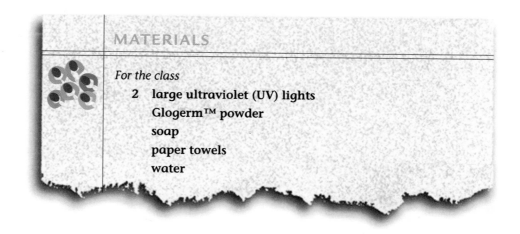

MATERIALS

For the class
2 large ultraviolet (UV) lights
Glogerm™ powder
soap
paper towels
water

PROCEDURE

Part One: Washing Your Hands

1. In your science notebook, make a table like the one shown below.

Table 1: Observations of Hands

	Hands Sprinkled With Powder	Hands That Were Shaken
Before		
After		

2. Have one person on your team of two students sprinkle a small amount of white powder on the palm of one hand. This person should spread the powder all over his or her hands by rubbing the hands together, covering the palms, backs of hands, fingers, and nails.

3. Firmly shake both hands with your partner. Do this by shaking right hand with right hand and left hand with left hand.

4. Look carefully at your hands and your partner's hands under the ultraviolet (UV) light. Record your observations in the first row of your data table.

5. Both you and your partner should wash your hands as you would *normally*.

6. Look carefully at your hands and your partner's hands under the UV light. Record your observations in Table 1.

Part Two: Improving Hand Washing

7. Design an experiment to improve the effectiveness of hand washing in removing microbes from the surface of your hands. For example, does the length of time you rub your hands make a difference? Is there a specific technique that is better for hand washing?

When designing your experiment, think about the following questions:

- What is the purpose of your experiment?
- What variable are you testing?
- What variables will you keep the same?
- What is your hypothesis?
- How many trials will you conduct?
- Will you collect qualitative and/or quantitative data? How will these data help you make a conclusion?
- How will you record these data?

8. Record your hypothesis and your planned experimental procedure in your science notebook.

9. Make a data table that has space for all the data you need to record. You will fill it in during your experiment.

10. Obtain your teacher's approval of your experiment.

11. Conduct your experiment and record your results.

ANALYSIS

Part One: Washing Your Hands

1. Where on your hands did you find the most "microbes" (white powder)?

2. Based on your results in Part One, how well did washing your hands remove "microbes"?

Part Two: Improving Hand Washing

3. Why is hand washing important? Use your knowledge of microbes and the results of this activity to explain your answer.

4. How well do powdered "microbes" model real microbes? Explain.

5. Imagine that your school has decided to launch a hand-washing campaign. You are in charge of designing the campaign and evaluating its effectiveness.

   a. Why might people resist changing the frequency and the way in which they wash their hands?

   b. Explain how you could persuade people to change their hand-washing behavior.

   c. What type of data could you collect (both before and after the campaign) to determine if the hand-washing campaign was effective?

6. Read the recommendations for hand washing for surgeons and food handlers on the next page. Why do both sets of guidelines stress rubbing or scrubbing the hands?

EXTENSION

Make a list of recommendations for a school hand-washing campaign. Explain how each recommendation would help reduce the spread of microbes.

Guidelines for Doctors Prior to Surgery

- Wet hands.

- Clean nails.

- Scrub hands (fronts and backs, each finger and between fingers) and forearms for 5 minutes, using antibacterial soap and a hand brush.

- Hold hands above elbow and allow excess water to drip off.

- Dry hands and forearms with a sterile towel.

- Put on surgical gown.

- Put on sterile gloves. (Many surgeons use double gloves.)

Guidelines for Food Industry Workers
(restaurant staff, supermarket workers, food packers, etc.)

- All personnel must wash their hands before returning to work.

- Wet hands with warm running water.

- Add soap, then rub hands together, making a soapy lather. Do this away from the running water for at least 15 seconds, being careful not to wash the lather away.

- Wash the front and back of hands, as well as between fingers and under nails.

- Rinse hands well under warm running water. Let the water run back into the sink, not down to your elbows.

- Turn off the water with a paper towel and dispose in a proper receptacle.

- Dry hands thoroughly with a clean towel.

ROLE PLAY

As you first learned in Activity 46, "Disease Fighters," your immune system recognizes and fights disease-causing microbes. Most people are able to fight off diseases like colds or the flu quickly and return to full health within a week or so. Other diseases, like diphtheria, are more severe. Such diseases are much more likely to have serious effects, or even lead to death, in a larger portion of the population.

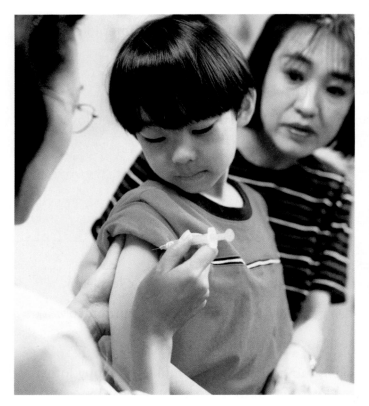

How can you fight serious diseases that often overwhelm human immune systems? One approach, **vaccination** (vak-suh-NAY-shun), is very effective in preventing some diseases. Each **vaccine** works against a specific disease. Vaccines are available against diseases caused by both viruses and bacteria. There are also vaccines being developed to work against other microbes as well.

CHALLENGE ⟩

How do vaccines prevent disease?

PROCEDURE

1. Assign a role for each person in your group. Assuming that there are four people in your group, each of you will read one role in Scene 1 and another role in Scene 2.

Roles in Scene 1	Roles in Scene 2
Student	Student
Parent	Parent
Older sibling (sister or brother)	Doctor
Grandparent	Nurse

2. Read the role play on the next pages aloud. Insert the names of your group members as directed.

ACT OR REACT!

SCENE 1: At the dinner table

Parent: It's getting to be flu season. My boss wants all employees to get flu shots. She doesn't want us to miss work and get behind on our deadlines.

Older Sibling: Funny you should mention that. My friend at school was just telling me that he has been feeling sick and thinks he has the flu. How does a flu shot stop you from getting sick?

Grandparent: A flu shot is a vaccine that helps your body resist the flu.

Student: Oh, I know about vaccines. I had a measles vaccine when I was a little kid and then you took me to get a booster shot just last year.

Parent: That reminds me—it's time for your tetanus booster.

Student: Another shot? I hate shots.

Parent: Yes, the immunity from some vaccines begins to wear off after a while, so you need a booster—it gives your immune system a boost.

Student: Why do I need to get all of these shots? And how does a vaccine work anyway?

➢

Older Sibling:	A vaccine is a dead or weakened form of a microbe or a part of the microbe. It helps your immune system prepare in advance to fight off the disease-causing germs.
Student:	You mean they actually inject you with the disease microbe?
Older Sibling:	Yup! The microbe is first inactivated, so it doesn't make you sick.
Student:	Do vaccines work only against diseases caused by viruses, like the flu?
Grandparent:	I don't think so. There are vaccines against tetanus and diphtheria, and they are caused by bacteria, not viruses. When my mother and father were very young, their parents worried that they would get tetanus, or "lockjaw," as they called it, every time they got a deep cut. They didn't have tetanus shots then.
Student:	So you're saying that all vaccines work to keep you from getting a disease, but not all diseases are caused by the same thing.
Parent:	Exactly. The way in which vaccines work is the same for different diseases, but each disease is caused by a different microbe, or germ. A microbe can be a bacterium or a virus.
Student:	Then what's the difference between bacteria and viruses?
Older Sibling:	This is exactly what we're studying in science class! Bacteria are living cells that grow and divide. Viruses can't grow or divide unless they inject their genetic material into another cell.
Grandparent:	(*Name of Older Sibling*), what do you mean by "genetic material"?
Older Sibling:	Oh, that just means DNA. You've probably heard of DNA.
Student:	I've heard of DNA. Viruses have DNA?
Parent:	Yes, I read about this. Viruses have a small amount of DNA, or sometimes a similar substance called RNA, as their genetic material. They also have an outer coat that protects the material inside, but that's about it. They infect cells and cause the infected cells to make new copies of the virus. But if a virus can't get into a living cell, no copies are made and the virus can't reproduce. That's why viruses aren't considered to be living organisms.
Student:	How do they make a vaccine? Why doesn't it make you sick?
Grandparent:	I don't know about all vaccines, but I do remember polio vaccines. Polio was a common disease when I was young. In fact, President Franklin D. Roosevelt became paralyzed from the waist down when he

caught polio as an adult. He hid from the public how serious it was, perhaps because of people's attitudes toward disabilities at the time.

Chemicals or heat were used to inactivate the polio virus and it was then injected into a healthy person. As (*Name of Older Sibling*) said, the inactivated virus didn't make you sick. But your immune system was tricked into getting ready to fight off an infectious polio virus.

President Franklin Delano Roosevelt

Parent: The polio vaccine is the only vaccine I know of that you can take orally. They used to put it on a sugar cube. The weakened virus wouldn't make you sick, but would still cause you to become immune.

Grandparent: I remember the first polio vaccine. It was a shot. My little brother hated it, but my mother was so relieved not to have to worry that we might get polio. It was a very serious threat back then. Vaccines have almost wiped out some diseases like polio and smallpox.

Student: Wow, until today, I never thought about life before vaccines.

SCENE 2: In the doctor's office

Nurse: Dr. (*Last name of Doctor*), here's your next patient.

Doctor: Hello (*Name of Student*), what seems to be the problem? Not feeling well?

Student: I think I have the flu.

Doctor: (checking pulse) Hmmm...let's see. What are your symptoms?

Student: Well, I've been coughing a lot and my throat's sore....I've been really tired. And I'm also starting to have a hard time breathing.

Parent: I think there's a fever, too.

➢

Doctor:	(to Student) Open your mouth and let's take a look.
	(Doctor examines Student's throat, listens to chest cough, and takes temperature.)
	Well, it could be the flu. But I'm not sure. You might have strep throat or pneumonia (new-MOW-nyah) or just a really bad cold. We'll have to take a chest x-ray and a throat culture.
Parent:	Are you going to prescribe an **antibiotic** (an-tih-by-AH-tik)?
Doctor:	Not yet.
Parent:	Why not?
Nurse:	Because if you have the flu, an antibiotic won't work. The flu is a viral disease, as are some types of pneumonia.
Parent:	I'm sorry, but I don't understand. I thought it was standard practice to prescribe an antibiotic. I always get one when I have the flu or a sore throat.
Doctor:	Nurse (*Name of Nurse*), why don't you explain while I give (*Name of Student*) a chest x-ray and throat culture?
	(Student leaves with doctor.)
Nurse:	Good idea. Antibiotics are medications that are used to fight bacterial diseases. This is because bacteria are living organisms that are killed by the action of an antibiotic.
Parent:	They always tell you to take the entire prescription.
Nurse:	Right. It takes time to kill the entire population of bacteria in your body. If you stop taking the antibiotic before all the disease-causing bacteria are gone, you run the risk of having them increase again.
Parent:	So why don't antibiotics work on diseases caused by viruses?
Nurse:	Because of the fact that viruses are not cells. They can reproduce only by entering your cells and using the cells to reproduce. That makes them harder to attack....Oh, here's (*Name of Student*), back from the x-ray room.
	(Doctor and Student return.)
Student:	Hey (*Mom/Dad*), my X-ray was negative, so I don't have pneumonia. I hope I don't have to take antibiotics. I hate swallowing pills!

Parent: Doctor, why don't you just go ahead and give us an antibiotic? The nurse just said that antibiotics are the way to treat bacterial diseases.

Doctor: That's right, but unless a throat culture or X-ray is positive, (*Name of Student*) probably doesn't have a bacterial infection. And it isn't helpful to take antibiotics that are not needed.

Student: But the last time I had the flu, you prescribed an antibiotic.

Doctor: That's right. But if I recall correctly, the last time you were here, your little sister was also sick—and she had a bacterial infection. You seemed to have the early symptoms of her infection. So the antibiotics were intended to treat the bacterial infection.

Student: (interrupting)...and not to cure the flu! So how do you cure the flu?

Nurse: There is no cure for the flu. Medicines you take when you have the flu or a cold only relieve the symptoms, like fever or headache. But they can't make you well; they can just make you feel better. In fact, there is no cure for most viral diseases.

Parent: What about flu shots?

Student: (*Mom/Dad*), you told me that flu shots prevent you from getting the flu. They don't make you better.

Doctor: That's right. Unfortunately, although flu shots prevent the flu in most cases, they don't work in every case. Why, just this morning we had a lady in the office who had gotten a flu shot, but who still caught the flu. It seems that she was infected with a different type of the flu than the one she had received a vaccine for.

Nurse: May I suggest that you take (*Name of Student*) home? We'll call you in the morning; by then we'll know if (*Name of Student*) needs an antibiotic.

Student: Yeah, I need to lie down. My head hurts from all this stuff. Maybe I'll ask my science teacher more when I feel better.

ANALYSIS

1. A vaccine prevents a person from catching an infectious disease; it does not treat the disease after the person has caught it. What are some advantages of preventing, rather than treating, infectious diseases?

2. Why are serious side effects from vaccines very rare?

3. You go to the doctor and find out that you may have the flu. Would you expect to be prescribed an antibiotic? Explain your answer.

4. Do you think that vaccinations against the flu should be required? Explain. Support your answer with evidence and identify the trade-offs of your decision.

   Hint: To write a complete answer, first state your opinion. Provide two or more pieces of evidence that support your opinion. Then consider all sides of the issue and identify the trade-offs of your decision.

5. **Reflection:** Explain whether you would change your answer to Question 4 if the disease had more severe symptoms and a greater chance of causing death.

EXTENSION

The vaccines for polio were developed in the 1950s. Find out more about how this disease affected society by asking different generations of your family, such as your parents and grandparents, if they can recall knowing anyone who had polio.

Despite many prevention methods, infectious diseases, from Hansen's disease to tuberculosis, continue to affect people in the U.S. and around the world. What can be done for a person after he or she has caught an infectious disease?

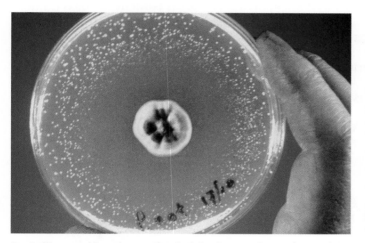

Penicillium *mold produces a chemical that has saved many human lives.*

Diseases caused by microorganisms, such as bacteria and protists, can usually be treated with antibiotics. This is because antibiotics are chemicals that kill living microbes such as bacteria. You may have heard of antibiotics such as penicillin (peh-nuh-SIH-lun) and streptomycin (strep-tuh-MY-sun). You may have even used them yourself. How were antibiotics first discovered? Did scientists design experiments and control variables? What problems did scientists face?

CHALLENGE

How was the first antibiotic discovered?

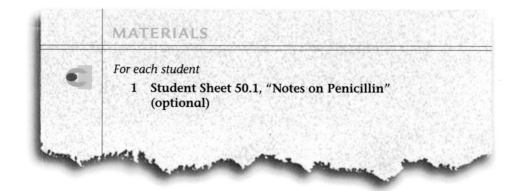

MATERIALS

For each student

1   Student Sheet 50.1, "Notes on Penicillin" (optional)

PROCEDURE

1. In order to prepare to watch the story on the video, first read Analysis Questions 1–3.

2. Find out more about antibiotics by watching a segment on the discovery of penicillin from the video, *A Science Odyssey:* "Matters of Life and Death."

3. Watch the video a second time and take notes on the following questions. Or use Student Sheet 50.1, "Notes on Penicillin," as a guide.

   • What was the scientific discovery? How was it made?

   • Who made it?

   • What was done as a result of this discovery?

4. Answer the Analysis Questions.

ANALYSIS

1. Describe the impact of penicillin on society.

2. Think back to the traditional scientific method, first discussed in Activity 1, "Solving Problems: Save Fred!" in Unit A, "Studying People Scientifically," of *Science and Life Issues*. Explain how the work of each of the following scientists did or did not resemble the traditional scientific method.

   a. Alexander Fleming

   b. Oxford University team (made up of 19 researchers, including Howard Florey and Ernst Chain)

3. What types of infectious diseases do antibiotics work against? Are there any types of infectious diseases that antibiotics do not work against? Explain.

MODELING

Have you ever taken antibiotics? Did you follow the directions completely? All antibiotics need to be taken as directed, which usually means taking all the pills and not stopping even if you begin feeling better. Why?

Millions of harmless bacteria naturally live on and inside of your body. When harmful bacteria appear on the scene, your body's immune system can usually keep a small population of them under control. If, however, these bacteria reproduce too quickly, you suffer consequences—and this is called an infection. Antibiotics help your body fight off an infection by killing these harmful bacteria. Unfortunately, a small number of bacteria in any population may not be affected by the antibiotic as quickly. These bacteria, which are considered more **resistant** to the treatment, continue to reproduce and grow. Completing the **full course** of the antibiotic as prescribed helps make sure that these bacteria do not survive and therefore won't make you ill or infect anyone else.

CHALLENGE

Why is it important to take an antibiotic as prescribed?

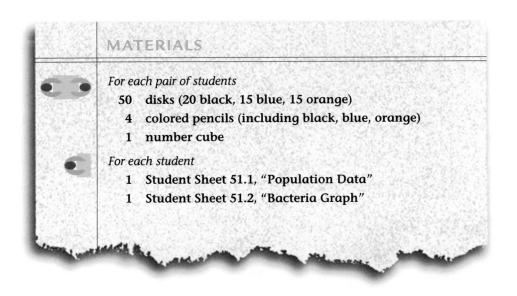

MATERIALS

For each pair of students

50 disks (20 black, 15 blue, 15 orange)

4 colored pencils (including black, blue, orange)

1 number cube

For each student

1 Student Sheet 51.1, "Population Data"

1 Student Sheet 51.2, "Bacteria Graph"

A BACTERIAL INFECTION

Imagine that you are sick with a bacterial infection. Your doctor prescribes an antibiotic to be taken every day for eight days.

Colored disks represent the harmful bacteria that are in your body:

Disease-Causing Bacteria	Represented By
Least resistant bacteria	black disks
Resistant bacteria	blue disks
Extremely resistant bacteria	orange disks

Each time you toss a number cube, it is time to take the antibiotic. The number on the number cube tells you what to do.

PROCEDURE

1. In this activity, you will work with your partner to collect data. Begin with 20 disks: 13 black, 6 blue, and 1 orange. These disks represent the harmful bacteria living in your body before you begin to take the antibiotic. Set the extra disks aside for now.

2. It is time to take your antibiotic. Toss a number cube and follow the directions in the Number Cube Key (on the next page).

Number Cube Key		
You Toss	**What Happened**	**What To Do**
1, 3, 5, 6	You took the antibiotic on time, so bacteria are being killed!	Remove 5 disks: remove all of the black disks first, then the blue, then the orange.
2, 4	You forgot to take the antibiotic.	Do nothing.

3. Record the number of each type of bacteria in your body in Table 1, "Number of Harmful Bacteria in Your Body," on Student Sheet 51.1, "Population Data."

4. *The bacteria are reproducing all of the time!* If one or more bacteria of a particular type are still alive in your body, add 1 disk of that color to your population.

   For example, if you have resistant (blue) and extremely resistant (orange) bacteria in your body, add 1 blue disk and 1 orange disk to your population.

5. Repeat Steps 2–4 until you have completed Table 1.

6. Use your data in Table 1 to graph the population for each type of bacteria and for the total number of bacteria on Student Sheet 51.2, "Bacteria Graph." Use different colored lines, or lines with different patterns, to represent each type of bacteria, and fill in the key accordingly.

ANALYSIS

1. Did the antibiotic help you to completely kill all of the harmful bacteria living in your body? Explain.

2. **a.** Imagine infecting someone else immediately after catching the infection (before you started taking the antibiotic). With what type of bacteria would you be most likely to infect them?

   b. Imagine infecting someone else near the end of your antibiotic course. With what type of bacteria would you be most likely to infect them?

   c. Suppose most infected people stopped taking the antibiotic when they began to feel better. (For example, consider the point in the simulation when there were only three harmful bacteria left.) What do you predict might happen to an antibiotic's ability to kill the harmful bacteria if the infection returns? Explain your reasoning.

3. Use your graph to describe how the population of each type of bacteria changed over the course of the antibiotic treatment.

4. Why is it important to complete the full course of an antibiotic as prescribed?

5. Was this activity a good model of an antibiotic treatment? Explain.

6. You find out that you have a viral infection and not a bacterial infection. What would happen to the amount of virus in your body each time you took the antibiotic? Explain.

TALKING IT OVER

In the last activity, you saw how important it is to follow directions and complete the full course of antibiotics as prescribed. Are antibiotics truly miracle drugs? Will they cure every infection? What can people do to maintain the effectiveness of antibiotics?

The Weekly Gazette
Antibiotic-Resistant Infection Strikes Local Hospital!

Metropolitan Post
When Miracle Drugs Fail

City Reporter
Bacteria Fight Back!

Evening Chronicle
Antibiotic Resistance on the Rise

The Daily Tribune
Are We Losing the Bacteria Wars?

CHALLENGE

What happens when antibiotics are overprescribed or used improperly?

PROCEDURE

1. Read about the miracle drugs known as antibiotics. As you read, think about what you might do if you were prescribed an antibiotic.

2. Discuss Analysis Question 1 with your group.

MIRACLE DRUGS—OR NOT?

What if someone told you that the pill you took to get better today might not work for you if you fall sick tomorrow? That's what health experts are saying about the miracle drugs known as antibiotics.

Common Antibiotics

Antibiotic	Brand Name	Used Against
Amoxicillin	Amoxil®, Polymox®, Wymox®, Trimox®	bronchitis, ear infections, sinus infections
Ampicillin	Unasyn®	urinary tract infections, meningitis
Cefaclor	Ceclor®	infections of the ear, nose, throat, respiratory tract, and urinary tract; strep throat; pneumonia; tonsillitis
Ceftriaxone	Rocephin®	Lyme disease, gonorrhea
Cephalexin	Keflex®, Keftab®	infections of the skin and urinary tract
Chloramphenicol	Chloromycetin®	typhiod, Rocky Mountain Spotted fever, meningitis
Clotrimazole	Lotrimin®, Mycelex®	yeast infections
Clindamycin	Cleocin®	pneumonia, strep throat, acne
Doxycycline	Atridox®, Doryx®, Doxy®, Periostat®, Vibramycin®	urinary tract infections, chlamydia, trichomonas
Erythromycin	Akne-Mycin®, EryDerm®, Erygel®, Ery-Tab®, Erythrocin®, Ilotycin®, Staticin®	Legionnaire's disease, pneumonia, strep throat, mild skin infections
Isoniazid	Nydrazid®	tuberculosis
Metronidazole	Flagyl®	amoebic dysentery, giardiasis
Monocycline	Minocin®	acne, amoebic dysentery, anthrax, cholera, plague, respiratory infections
Mupirocin	Bactroban®	skin infections, impetigo
Penicillin	various	strep throat, pneumonia, syphilis, dental and heart infections
Tetracycline	Achromycin®	respiratory infections, pink eye, pneumonia, severe acne, typhoid, Rocky Mountain Spotted fever

Colony of bacteria. Some microbes are antibiotic-resistant (orange).

Use of antibiotics kills majority of bacteria, except those that are antibiotic-resistant.

Without competition, antibiotic-resistant bacteria increase.

Figure 1: The Rise of Antibiotic-Resistant Bacteria

Antibiotics have been used to fight diseases for over 50 years. Today, they are losing their effectiveness. This is the result of more antibiotic-resistant bacteria. In the last 10–15 years, antibiotic-resistant bacteria have included strains of *Mycobacterium tuberculosis*, which causes tuberculosis (TB), and *Streptococcus pneumoniae*, the most common cause of human ear and sinus infections.

Shown on page C-104 are some common antibiotics. Do you recognize any antibiotics that you have taken?

Reasons for the development of antibiotic-resistant bacteria include overprescription and incorrect use of antibiotics. "Using antibiotics incorrectly has led to the development of bacteria that can resist them," says Dr. Richard Dietrich of Kaiser Permanente in Baltimore, Maryland. This means that the drugs people rely on to cure everything from strep throat to bacterial pneumonia may not work when they are taken.

Most antibiotics must be taken over a period of time. When patients feel better, they sometimes stop taking the medication and don't complete the full course of treatment. In such cases, antibiotic-resistant bacteria may not be killed by the medication. They are more likely to reproduce and grow without competition from other microbes that have been killed by the drug. If the antibiotic-resistant bacteria cause disease, it becomes difficult to treat the patient with antibiotics (see Figure 1).

"The bacteria that cause pneumonia and ear and sinus infections commonly live in our throats and noses," Dr. Dietrich says. "If you take an antibiotic for no good reason, it kills only the germs that are not resistant to the antibiotic. An infection caused by the remaining resistant bacteria can be very hard to treat." If you take antibiotics when you don't need them, the drugs may lose their ability to help you get better when you really do need them.

One reason antibiotics are overused is that so many patients ask for them. Dr. Dietrich adds that it's common for patients to believe that antibiotics will cure whatever illness they have. But antibiotics work against only certain microbes, such as bacteria. They do not work against viruses. Doctors also used to prescribe antibiotics more often, partly as a precaution against disease. Now, this is less common and antibiotics are prescribed only for specific diseases.

Adapted with permission from Kaiser Permanente, 2001

For links to more information on antibiotic resistance, go to the SALI page of the SEPUP website.

ANALYSIS

1. Describe what can happen if people take antibiotics when they don't need them.

2. What is one reason antibiotics are overused?

3. You have a sore throat and there are some antibiotics left over from your brother's strep infection last month. Should you take them for your sore throat? Why or why not?

4. Your friend is prescribed an antibiotic on Monday. Suppose she feels better two days later. Should she stop taking the medicine? Explain.

5. **Reflection:** Think about what you have learned in the last few activities. Imagine you don't feel well and the doctor tells you that you have the flu. The doctor suggests taking an antibiotic. What would you do?

EXTENSION

Design a survey to find out what people know about the correct use of antibiotics. Good survey questions should be clear (the person should know exactly what you are asking) and concise (ask for only one piece of information per question). To make it easy to quantitatively analyze the survey data, develop questions that can be answered either yes or no. Examples of good questions include: When you are prescribed an antibiotic, do you always take all of it? Do you always expect a doctor to prescribe an antibiotic when you are sick?

INVESTIGATION

Ebola fever, Lassa fever, Hanta virus, Bolivian hemorrhagic fever, and AIDS are all examples of new infectious diseases. Some new infectious diseases are the result of new interactions between people and the environment. Many of these diseases can be traced to animal species. For example, when people go deeper into unexplored jungles, they are more likely to come into close contact with wild animals and their diseases, perhaps for the first time. In some cases, the disease passes from animal to human. Ebola is probably one such disease. Epidemiologists believe that Ebola originally may have been an infection of green monkeys in Uganda.

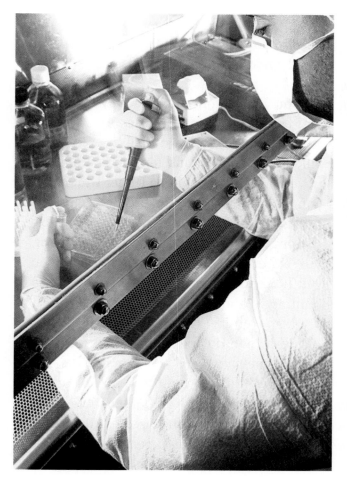

There is a real risk that such new diseases could quickly spread among different populations anywhere in the world. That is why the United States is prepared to send scientific teams to respond immediately to possible outbreaks of these new diseases, which are also known as **emerging diseases**.

In this activity, you will simulate the experience of a team of epidemiologists trying to trace the cause of a new disease.

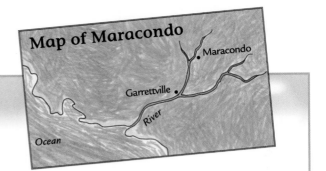

Map of Maracondo

MARACONDO FEVER

An old man struggled out of a canoe in the tropical heat and stumbled into the town of Garrettville, collapsing in the main street. The children who found him were shocked to see that his eyes were bloodshot, his nose was bleeding, and his skin was bruised. He was immediately rushed to the hospital where he began coughing up blood. Before becoming unconscious, he told the doctors of a frightening disease affecting the people of his village, Maracondo. The old man died later that day of the same disease he had been warning the townspeople about.

None of the medical staff knew much about Maracondo (population 85), which is the last village that boats can reach as they head up the river. The medical staff of the Garrettville Hospital was worried about the spread of this mysterious disease. No one in the town became ill, but the townspeople were very frightened. They were also concerned about the people in Maracondo. The hospital doctors collected samples from the dead man, including urine, blood, and feces. They sent these samples, packed in dry ice, to the United States—to the Centers for Disease Control and Prevention (CDC) in Atlanta, Georgia. When the samples were analyzed, it appeared that the blood of the man contained an unknown virus.

The CDC worked quickly to put together an expert team to help the people of Maracondo. A doctor, an epidemiologist, a veterinarian, and an ecologist make up the team. The team's mission is to go to Maracondo to find out how this new disease is spread in order to stop more people from getting sick. If the illness is spread from person to person, it might take only one infected person to get on a plane to accidentally start an epidemic more horrifying and deadly than the bubonic plague of the 1300s.

• • •

You and your partners make up the expert team from the CDC. You must gather evidence to determine how this disease, now called Maracondo Fever, is transmitted. You need to do this in time to save the people of Maracondo.

As you begin your investigation, imagine you are heading into Maracondo. You are very hot, thirsty, and irritable. People in the village are terrified and tension is high. You have no idea if the village leader who meets you has the virus in his breath, or on his hands, or if the virus is being carried by mosquitoes that are, at this very moment, buzzing around your head. In addition, you will have to live in a straw hut, sleep in a hammock, and boil all of your water.

But as a "can do" person, you get local villagers to help you set up a lab in half the normal time. You also build an animal collection center to check if any of the local animals are carrying the disease. All the people on your team are experienced microbe hunters and know that hunches are not good enough. You owe it to the villagers and the rest of the world to base your conclusions and recommendations on strong evidence.

CHALLENGE

How is Maracondo Fever spread? What can you do to stop it?

MATERIALS

For each group of four students

1 Maracondo Fever game board
1 set of Hut, House, Lab, and Weird Events cards
1 number cube
4 game pieces
1 Student Sheet 53.1, "Maracondo Fever Hypotheses"
1 Student Sheet 53.2, "Infected People"
1 Student Sheet 53.3, "Healthy People"
1 Student Sheet 53.4, "Additional Field Notes"

PROCEDURE

Part One: Field Notes

1. Before arriving in Maracondo, discuss with your team the symptoms and possible causes of the disease. Brainstorm ways in which the disease may be passed around the community.

2. After arriving at Maracondo, you gather additional information. Review your field notes (on the next page) before going on to Step 3.

3. How do you think Maracondo Fever is spread? Discuss your ideas in your group and write out your initial hypothesis on Student Sheet 53.1, "Maracondo Fever Hypotheses."

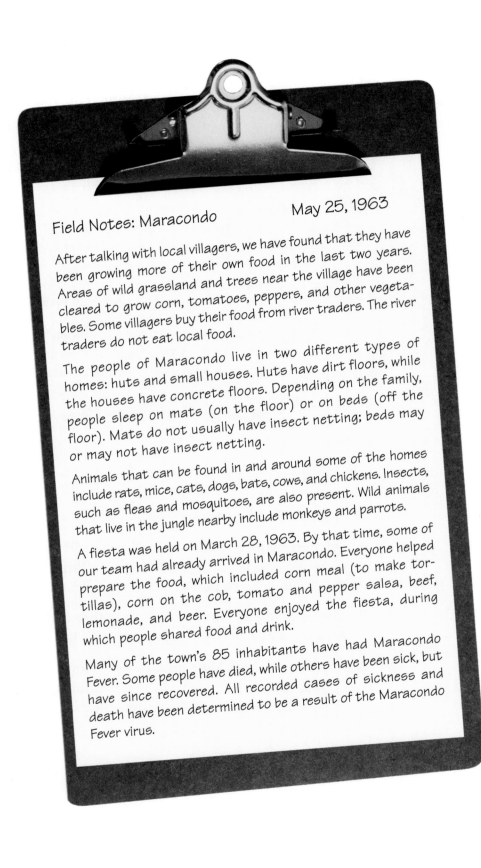

Field Notes: Maracondo   May 25, 1963

After talking with local villagers, we have found that they have been growing more of their own food in the last two years. Areas of wild grassland and trees near the village have been cleared to grow corn, tomatoes, peppers, and other vegetables. Some villagers buy their food from river traders. The river traders do not eat local food.

The people of Maracondo live in two different types of homes: huts and small houses. Huts have dirt floors, while the houses have concrete floors. Depending on the family, people sleep on mats (on the floor) or on beds (off the floor). Mats do not usually have insect netting; beds may or may not have insect netting.

Animals that can be found in and around some of the homes include rats, mice, cats, dogs, bats, cows, and chickens. Insects, such as fleas and mosquitoes, are also present. Wild animals that live in the jungle nearby include monkeys and parrots.

A fiesta was held on March 28, 1963. By that time, some of our team had already arrived in Maracondo. Everyone helped prepare the food, which included corn meal (to make tortillas), corn on the cob, tomato and pepper salsa, beef, lemonade, and beer. Everyone enjoyed the fiesta, during which people shared food and drink.

Many of the town's 85 inhabitants have had Maracondo Fever. Some people have died, while others have been sick, but have since recovered. All recorded cases of sickness and death have been determined to be a result of the Maracondo Fever virus.

Part Two: Maracondo Fever Game

4. Begin the game by placing the cards face down in four stacks: Hut, House, Lab, and Weird Events. You will pick up a card every time you land on a space. For example, if you land on a Hut, read a Hut Card. Do the same for House, Lab, and Weird Events spaces.

5. Each person begins on the Start space. Have one person from the team toss the number cube and move that number of spaces on the game board.

6. Pick up and read the card. As a team, record the information you learn on

- Student Sheet 53.2, "Infected People"

- Student Sheet 53.3, "Healthy People"

- Student Sheet 53.4, "Additional Field Notes"

7. As you gather more evidence, revise your hypothesis on Student Sheet 53.1, "Maracondo Fever Hypotheses." When you have too much evidence against a hypothesis, develop another hypothesis that fits the evidence.

8. Have the next player toss the number cube and move his or her game piece. As a team, repeat Steps 6 and 7.

9. Continue playing and collecting evidence. When the first person passes the Start space, pause to have a team meeting. Discuss how you think the disease was spread in light of the new evidence you have collected. Be sure to record your revised hypothesis on Student Sheet 53.1.

10. Continue playing and collecting evidence. When the next person passes the Start space, turn over all of the remaining cards and record all of the evidence.

11. As a group, complete Analysis Questions 1 and 2.

ANALYSIS

1. **a.** Review your data on Student Sheet 53.2. What did the people who were infected have in common?

   b. Review your data on Student Sheet 53.3. What did the people who remained healthy have in common?

   c. Compare the data from Student Sheet 53.2 with the data on Student Sheet 53.3. What are some of the differences between those people who became infected compared with those who stayed healthy?

2. **a.** How do you think people are infected with Maracondo Fever? Explain how your evidence supports your final hypothesis.

   b. *People's lives are at stake!* Identify any evidence that seems to conflict with your final hypothesis and explain how your hypothesis addresses it.

3. Now that your CDC team has discovered how this disease spreads, you must recommend ways to reduce the spread of the disease, both within and outside of Maracondo. Recall what you know about viruses, as well as the information provided in this activity. Provide at least two recommendations to stop the spread of Maracondo Fever. Support them with evidence and identify the trade-offs involved in your recommendations.

   Hint: To write a complete answer, first state a recommendation. Provide two or more pieces of evidence that support your recommendation. Then consider the possible consequences of your recommendation and identify the trade-offs of your recommendation.

4. **Reflection:** What character traits and habits of mind would make a great epidemiologist?

Our Genes, Our Selves

Unit D

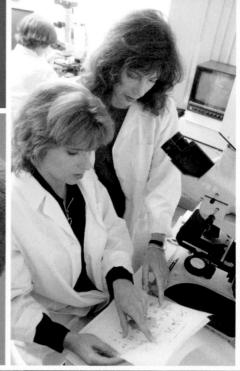

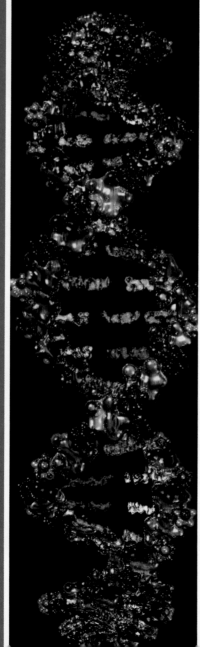

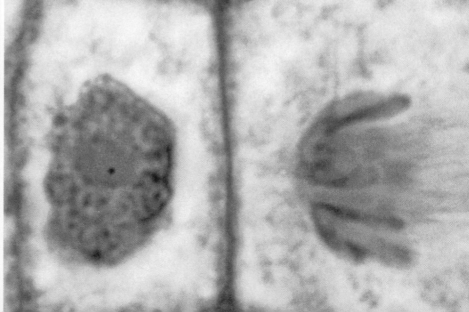

Our Genes, Our Selves

Grace thought her new brother would look more like her than her adopted sister does. But as she looked down at the new baby, she just couldn't see a resemblance.

"Look at his light eyes and red hair! No one else in our family has red hair," Grace remarked.

Her mother smiled. "Well, Dad's sister had red hair when she was young. Then it got darker."

Grace was surprised to hear that. But she still didn't understand why her baby brother looked so different from her.

• • •

Have you ever wondered why some children look very much like their biological parents while others look completely different? Why doesn't a child look like a simple blend of his or her parents? Why do some siblings look so different?

In this unit, you will begin to find answers to these questions. Because many of the same principles apply to all organisms, you will study humans and other organisms as well.

Some children look very similar to one of their biological parents. Some appear to be more of a blend of both parents, while others don't look very much like either parent. What are the reasons for this variation in family resemblance? What causes variation among people in general? You will look at six different human **characteristics**, such as eye color, to study human variation. Each of these characteristics can occur in different versions, or **traits** (TRATES).

CHALLENGE

How much variation is shown by the students in your class?

MATERIALS

For each pair of students

1   meter stick, tape measure, or height chart

For each student

1   Student Sheet 54.1, "Human Traits: Group Results"

1   Student Sheet 54.2, "Human Traits: Class Results"

1   piece of PTC paper

1   piece of control paper

1   sheet of graph paper

PROCEDURE

Figure 1: Tongue Rolling

1. Working with your group, decide whether each person's eyes are blue, gray, green, brown, or hazel (hazel eyes are a very light brown with yellow or green tones). If a person's eyes are difficult to classify, choose the color that is closest. Record your results on Student Sheet 54.1, "Human Traits: Group Results."

2. Try to roll your own tongue into a U-shape similar to that shown in Figure 1. On the student sheet, record who can and who cannot roll his or her tongue.

3. Try to cross all the fingers of the hand you normally write with as shown in Figure 2. You may use your other hand to help position the fingers. You should begin by crossing your pointer finger over your thumb, then try to cross your middle finger over your pointer finger. Continue trying to cross each finger, one by one, on top of the next finger. On the student sheet, record who can and who cannot cross his or her fingers like this.

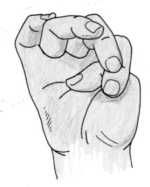

Figure 2: Finger Crossing

4. Working with a partner, use a meter stick or height chart to measure each other's height in centimeters (cm). Round to the nearest 5 cm and record the results on the student sheet.

5. Working with a partner, use a meter stick or measuring tape to measure each other's arm span in centimeters (cm). Obtain the arm span by spreading your arms out sideways as far as possible, and having your partner measure from the tips of the fingers on one hand to the tips of the fingers on the other hand, as shown in Figure 3. You may have to ask another student to help you hold the meter stick or measuring tape. Round to the nearest 5 cm and record the results on the student sheet.

Figure 3: Measuring Arm Span

6. Obtain one piece of plain paper and one piece of PTC paper from your teacher.

    a. Place the piece of plain paper on your tongue and move it around to be sure it mixes with your saliva. Then remove and discard the piece of paper as directed by your teacher.

    b. Do the same thing with the PTC paper. Record whether the PTC paper tastes different from the plain paper. If it tastes different, also record whether the taste is mild or strong.

7. Complete Table 1, "Group Results Summary," on Student Sheet 54.1. Note that you do not have to record totals for the height and armspan.

8. Have one person report your group's results to your teacher.

9. Record the class totals on Student Sheet 54.2, "Human Traits: Class Results."

10. Prepare a bar graph of the class data of one of the traits, as assigned by your teacher. Be sure to label your axes and title your graph.

ANALYSIS

1. For each of the six characteristics you studied, how many versions, or traits, are observed in your class? You can answer this question by completing the table below. (For example, if your class has people with brown and blue eyes only, then you would fill in the first column with "eye color," the second column with "brown and blue," and the third column with the number "2" to represent the two colors observed.)

Characteristic	Traits	Number of Traits
Eye color		
Tongue rolling		
Finger crossing		
PTC tasting		
Height		
Arm span		

2. Which of the traits you investigated—for eye color, tongue rolling, PTC tasting, crossing all your fingers, height, and arm span—do you think people inherit from their biological parents? Explain.

3. If a trait is not inherited, what else might cause it? Explain or give some examples.

4. If you studied more people in your community, would you expect to find more traits for each characteristic? Explain your answer.

EXTENSION

Gather data on ten more people who are not in your class and bring the results to class to add to the totals.

LABORATORY

Some traits appear to be passed from parents to their children. These are called **inherited** traits. The inherited bits of information that are passed directly from the parents' cells to the child's cells are called **genes**.

How do scientists find out whether a specific trait is inherited? If a trait is inherited, what information can the pattern of inheritance provide? Can anything else affect an inherited trait, or do genes determine everything about you?

Flowering Nicotiana *plants*

All organisms have genes and inherited traits. Scientists have learned a lot about how traits are inherited by studying other organisms. Scientists who study genetics often investigate yeast, plants, fruit flies, or other organisms that reproduce quickly. Through these studies, they have made discoveries that also apply to humans.

You will investigate plants to learn more about how traits are inherited. You will look at the colors of the plants right after they first sprout from seeds. The seeds will be from *Nicotiana*, a garden plant.

CHALLENGE

What color leaves will you observe on the offspring of two green parent plants?

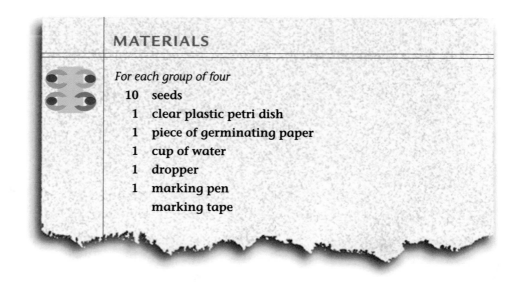

PROCEDURE

1. Observe the pictures of plant seedlings shown in Figure 1. The pictures show seedlings of the parent and grandparent plants that produced the seeds you will plant. In other words, the seeds you will sprout are offspring of the parent plants displayed in the figure. Think about the possible colors of the offspring that may grow from the seeds you plant.

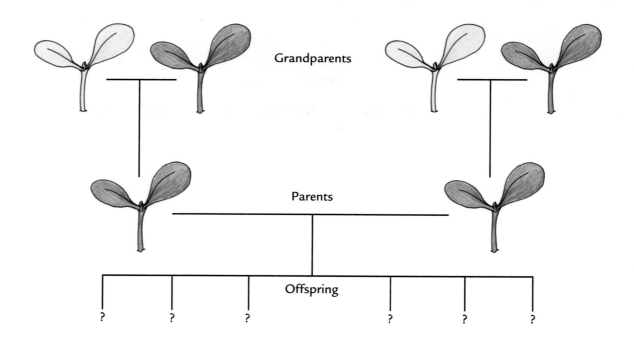

Figure 1: Plant Generations

2. Place a piece of germinating paper into the bottom of your petri dish.

3. Fill the dropper with water and add drops to the paper until the paper is wet. Pour off any excess water, so that the seeds will not drown.

4. Arrange 10 seeds in the dish. Try to leave plenty of room between each seed so they will have room to grow.

5. Put the cover on the dish and use tape and a marking pen to label the dish with your group members' names. Place the label near the side of the dish.

6. Following your teacher's directions, place the dish where the seeds will receive plenty of light.

7. Check your seeds every day or two. Carefully add a few drops of water to each piece of germinating paper as needed to keep the papers moist. Avoid having any excess water in the dishes. In about ten days, your seeds will have sprouted and grown enough for you to observe their appearance. At that time, you will complete Activity 62, "Analyzing Genetic Data."

ANALYSIS

1. Record in your science notebook your prediction for the color or colors of the plants that will grow from the seeds. You may make more than one prediction, but be sure to indicate which you think is most likely to happen.

2. What are your reasons for each prediction you proposed for Question 1? Explain.

VIEW AND REFLECT

About 100 years ago, a French doctor named Antoine Marfan described the symptoms of a young patient. Today, those symptoms are recognized as a genetic disease known as the Marfan syndrome. A syndrome is a condition that causes a pattern of physical changes. The physical changes observed in the Marfan syndrome are a result of changes in the body's connective tissue.

Connective tissue provides the connections between tissues in the body. It is found in all your organs. For example, if you ever looked closely at uncooked chicken, you may have noticed the connective tissue that attaches the skin to the muscle of the chicken.

Today, genetic counselors advise people who suspect they may have a genetic condition. If one member of this couple has the Marfan syndrome, what are the chances that any child they have will have the Marfan syndrome?

CHALLENGE

Would you want to find out if you could have a genetic disease? Why or why not?

PROCEDURE

1. Read the e-mail below from Joe to his friend Megan.

2. Prepare a list in your science notebook of the questions you think Joe should ask Dr. Foster before he decides whether to be tested.

3. Make a table in your science notebook to record the advantages and disadvantages of being tested, based on what you know so far about the Marfan syndrome. Leave room to add more rows to your table later.

4. In order to prepare to watch the story on the video, first read Analysis Questions 1–4.

5. Find out more about the Marfan syndrome by watching the video "How Do Your Genes Fit?" Look for answers to Analysis Questions 1–4 and to your own questions from Step 2 as you watch the video.

To: meganR@talk.com

From: joeF@email.com

Hey Megan—

I'm trying to act cheerful because Dad says you feel the way you act, but I'm feeling kind of down. I miss my mom. It's been exactly three years since she died of that strange heart condition. The doctor didn't even know about the condition until she died.

Now Dr. Foster is saying he thinks it might have been genetic, and that means I could have inherited it. He went to a medical convention where he learned more about a condition called the Marfan syndrome. He began to suspect it caused Mom's heart problem. Now Dr. Foster's

➢

saying I should be tested even though I'm totally healthy. He says the test will tell if I inherited this Marfan gene from my mom.

He says some of the symptoms of this syndrome are being very tall and having a long face and loose joints. I'm tall and have a long face, like my mom, but I'm not double-jointed as she was. She could bend her fingers in the strangest positions—my sister and I always thought it was so cool!

They say there's a way to tell if I've got the gene my mom had, but it's complicated. Lots of our relatives would have to give blood to be tested too. Dad doesn't want me to have the test. He doesn't know if our health insurance will cover the costs. He says that doctors are making too much of a fuss about genetic diseases, and says we're better off just taking what comes and not knowing too much. Then he says if I have the gene, we might lose my health insurance and people will find out and won't hire me when I'm older.

But I keep thinking about some things Dr. Foster told me. He told me about a volleyball player named Flo Hyman who was in great shape, but then died suddenly in the middle of a game because she had this Marfan syndrome. Then I saw in the newspaper they think that the guy who wrote the musical *Rent* might also have had it. Dr. F. says if I know, I can be careful and maybe have surgery to prevent heart problems. But what if I had to give up playing soccer?

I sure wish your mom hadn't gotten that new job and moved you halfway across the state. What do you think I should do? Should I talk Dad into letting me have the test? Or should I try to forget the whole thing?

Miss ya,

Joe

ANALYSIS

1. What are the signs that suggest a person may have the Marfan syndrome?

2. What causes the Marfan syndrome?

3. Can you "catch" the Marfan syndrome from another person, the way you can catch the flu? Explain.

4. What effect can the Marfan syndrome have on a person's life?

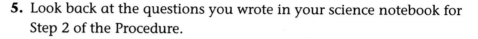

5. Look back at the questions you wrote in your science notebook for Step 2 of the Procedure.

   a. Were any of your questions answered? Record the new information you learned from the video.

   b. What new questions would you want to ask a doctor or genetic counselor?

6. **Reflection:** How would you behave toward a fellow student whom everyone suspects has the Marfan syndrome?

EXTENSION

Go to the SALI page of the SEPUP website for links to websites on the Marfan syndrome and other genetic conditions.

READING

Every human is the child of two biological parents. Most organisms that you are familiar with, like the kittens in the photo, are also the offspring of two parents. To be a "copycat" is to imitate someone as perfectly as possible. However, kittens and human infants are not exact copies of their parents. To understand why that is, it helps to start by looking at organisms that *do* copy themselves.

CHALLENGE

What is the difference between sexual and asexual reproduction?

READING

Asexual Reproduction

If you go to an art gallery, you might see an art historian inspecting a work of art, trying to determine whether it is the original or a "reproduction." A reproduction may not be a perfect copy of the original, but it's close.

Organisms reproduce in two ways. Some organisms can use **asexual reproduction** (in which they make exact copies of themselves). If you completed the Micro-Life unit of *Science and Life Issues*, you learned about bacteria and other microorganisms. These organisms reproduce themselves asexually. A single-celled organism, such as a bacterium or an amoeba, reproduces by dividing in two (see Figure 1). Each new organism produced (offspring) is identical to the parent cell.

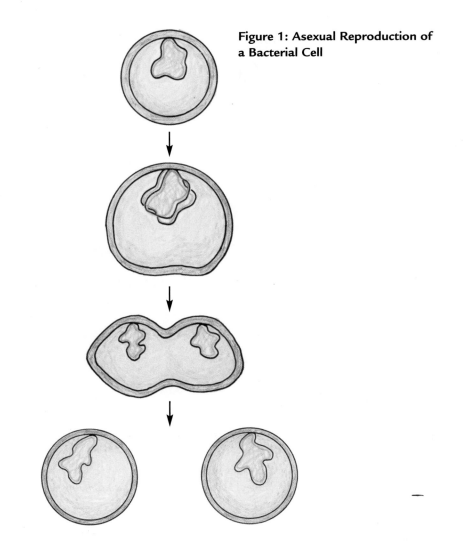

Figure 1: Asexual Reproduction of a Bacterial Cell

STOPPING TO THINK 1

In asexual reproduction of a bacterial cell, is it clear which cell is the parent and which is the offspring? Explain.

These two identical offspring are both called clones of the single parent organism. A clone inherits all of its traits from its one parent. If you completed the Micro-Life unit, you may remember the clumps of bacteria that grew in the petri dishes in Activity 47, "Reducing Risk." Each clump was formed by many divisions over time, starting from one bacterial cell. In other words, each bacterium in the clump is a clone reproduced from that one original cell (see Figure 2). Each of these clones has identical traits, except in rare cases when a random change occurs. Imagine copying a sentence and making a small mistake. In the same way, a gene can be reproduced slightly differently from the original gene. This change can cause the offspring cell to have a different trait from the parent cell. This type of random change is known as a **mutation** (myoo-TAY-shun).

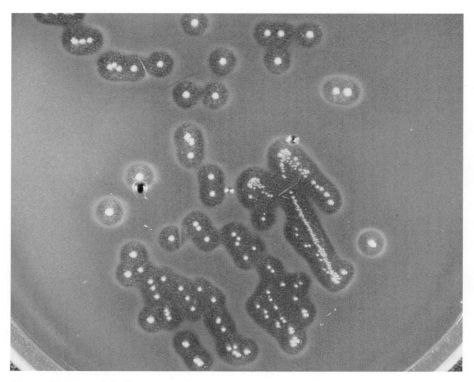

Figure 2: Bacterial Clones
Each clump of bacteria on this petri dish contains many cells, each identical to an original parent cell.

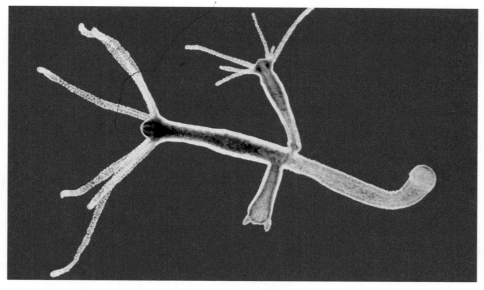

Figure 3: Budding Hydra
One way the Hydra reproduces itself is by asexual budding. The "buds" will soon break off from the parent to become identical offspring organisms.

Even some multicellular organisms can reproduce asexually. Budding is the name given to a process in which a small new organism grows directly out from the parent's body. One animal that reproduces this way is the *Hydra* (see Figure 3).

The strawberry plant (see Figure 4) can also reproduce asexually, by generating tiny new plants on a rootlike runner. Each of these little plants can eventually separate from the parent and become a new individual identical to its parent. Any organism that is produced through asexual reproduction can be considered a clone, since it inherits all its traits from one parent.

runners

parent plant

Figure 4: Asexual Reproduction of a Strawberry Plant
A strawberry plant can produce a new copy of itself by using runners.

new plant

STOPPING TO THINK 2

Your friend tells you, "Only single-celled organisms reproduce asexually. After all, how could a multicellular organism do that?" How do you respond to your friend?

Sexual Reproduction

Most animals and plants can also reproduce by **sexual reproduction**. In fact, humans and many other animals can only reproduce sexually. Such organisms inherit traits from two parents, not one. You are not a perfect copy of either of your parents. You also are not a perfect blend of your parents' traits. What determines which traits you get from each parent?

Sexual reproduction occurs when a tiny sperm cell produced by a male unites with an egg cell produced by a female, as shown in Figure 5. The union of the sperm and the egg is called **fertilization** (fur-tul-uh-ZAY-shun).

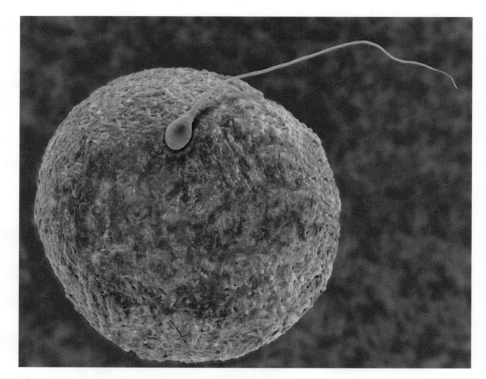

Figure 5: Sperm and Egg: Fertilization
The tiny sperm cell is about to fertilize a large egg cell.

The cell that results will have genes from both the egg and the sperm. Because genes help to determine traits, sexual reproduction produces a new cell that develops into an individual with traits inherited from both parents. This new individual will be different from each of its biological parents. Also, because no two sperm or egg cells contain exactly the same information, no two offspring produced by the same parents are identical. A unique set of inherited genes means a unique set of inherited traits.

There is an exception to the last statement: identical twins. Identical twins result when one fertilized egg splits before beginning to grow and develop—the two eggs then become two genetically identical offspring or children. Only one fertilization occurs, but two organisms are produced.

Asexual and sexual reproduction are summed up in the table below.

Reproduction	
Asexual	**Sexual**
One parent	Two parents
Offspring (clones) identical to parent	Offspring inherit traits from both parents

STOPPING TO THINK 3

"Fraternal twins" result when two eggs are both fertilized by sperm cells, and both develop into offspring. (This is very common in dogs and cats.) Why are identical twins much more similar than fraternal twins? Explain.

Cloning

In nature, a clone is an offspring produced by asexual reproduction. Yet you've probably heard the term cloning used to describe a process that produces a sheep (or other animal) identical to its one parent. How can this be possible given that sheep (like humans) always have two parents? (Remember, mammals never reproduce asexually.)

Scientists have now managed to perform asexual reproduction artificially. Consider the cloning of a sheep, illustrated in Figure 6 below. The part of the cell that contains the genes (the nucleus) is removed from one of the cells of the sheep's body. This nucleus is placed into an egg cell whose nucleus has already been removed. If this artificially fertilized egg develops in a womb, it grows into a sheep that is genetically identical to the sheep that its nucleus came from. It is therefore a clone of that sheep.

egg cell from another sheep with nucleus removed

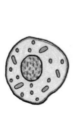

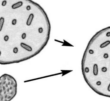

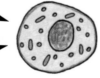

body cell from sheep

nucleus from body cell

nucleus is injected into egg cell

adult sheep to be cloned

artificially fertilized egg develops into a (younger) clone of first sheep

Figure 6: Cloning a Sheep

STOPPING TO THINK 4

How is a clone different from an identical twin?

ANALYSIS

1. Classify each of the following as either sexual or asexual reproduction. Explain each answer.

   a. An orange cat is mated with a black cat, in hopes of producing a tortoiseshell cat.

   b. A cutting is taken from a red-flowered geranium and placed in water to develop roots. Once roots have grown, the new plant is placed in soil and grows to produce another red-flowered geranium.

   c. A red-flowered geranium with dull leaves is bred with a white-flowered geranium with shiny leaves, with a goal of producing a red-flowered geranium with shiny leaves.

   d. A male fish releases sperm cells into the water. One of the sperm cells unites with an egg from a female fish to form a new cell that grows into a new fish.

   e. A small worm that lives in water splits in two and each half grows to normal size. The head end grows a tail, and the tail end grows a head.

   f. Sheep reproduce only by sexual reproduction in nature. Using modern technology, a clone of an adult sheep is produced.

2. **Reflection:** If you were given an opportunity to clone yourself, would you do it? Explain.

MODELING

Genes determine inherited traits by carrying the information that is passed from parents to offspring. These genes carry information that each cell of an organism needs in order to grow and perform its activities. How are genes for a trait passed from parents to offspring? How do they determine the offspring's traits?

CHALLENGE

How are simple inherited traits passed from parents to their offspring and then to the next generation?

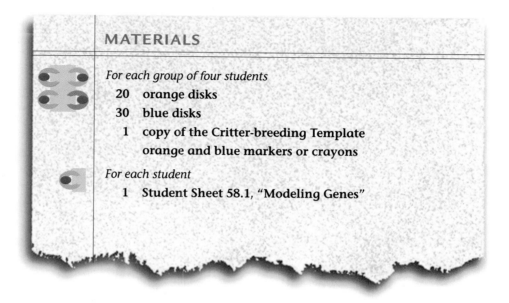

MATERIALS

For each group of four students
20   orange disks
30   blue disks
 1   copy of the Critter-breeding Template
     orange and blue markers or crayons

For each student
 1   Student Sheet 58.1, "Modeling Genes"

PROCEDURE

1. Work with your group to read and discuss the story that follows.

2. In order to evaluate your ideas, use Student Sheet 58.1, "Modeling Genes," to model the behavior of genes passed from parents to offspring.

ENDANGERED CREATURES

Part One: The First Generation

Imagine two islands in the ocean, far from land. The only known population of blue-tailed critters lives on one island. The only known population of orange-tailed critters lives on the other island. The critter population is shrinking and critters have just been classified as an endangered species. Although they produce many offspring, very few of the critter offspring survive in the wild because most are eaten by the black-billed yellowbird.

Skye is a rare blue-tailed critter.   Poppy is a rare orange-tailed critter.

Critters are hard to capture, so very few critters exist in captivity. Skye, a blue-tailed critter, lives in the Petropolis zoo. Poppy, an orange-tailed critter, lives in the Lawrenceville zoo. Critters reproduce by sexual reproduction. The research departments in these two world-famous zoos have decided to try breeding Skye and Poppy in order to produce offspring and keep the rare critters from becoming extinct.

STOPPING TO THINK 1

What do you think the tails of Skye and Poppy's offspring will look like? Explain your opinions to your group.

Part Two: The Second Generation

The breeding program is a great success. Skye and Poppy produce 100 offspring!

However, all 100 of these second-generation critters have blue tails. The scientists are concerned. "Will the orange-tail trait be lost?" they wonder.

The zoo scientists wonder why none of the critter pups have orange tails. They begin to argue about several possible explanations.

STOPPING TO THINK 2

Discuss this question with your group: Why do all of the offspring have blue tails? Develop one or more hypotheses. Be prepared to share one of your hypotheses with the class.

➤

Part Three: The Third Generation

Further attempts to breed Skye and Poppy are unsuccessful. However, once Skye and Poppy's offspring mature, they begin to have pups of their own. The scientists are fascinated by the results. Some of Skye and Poppy's "grandpups" have orange tails. The scientists notice that about one-fourth of all the pups in this third generation have orange tails. The rest have blue tails.

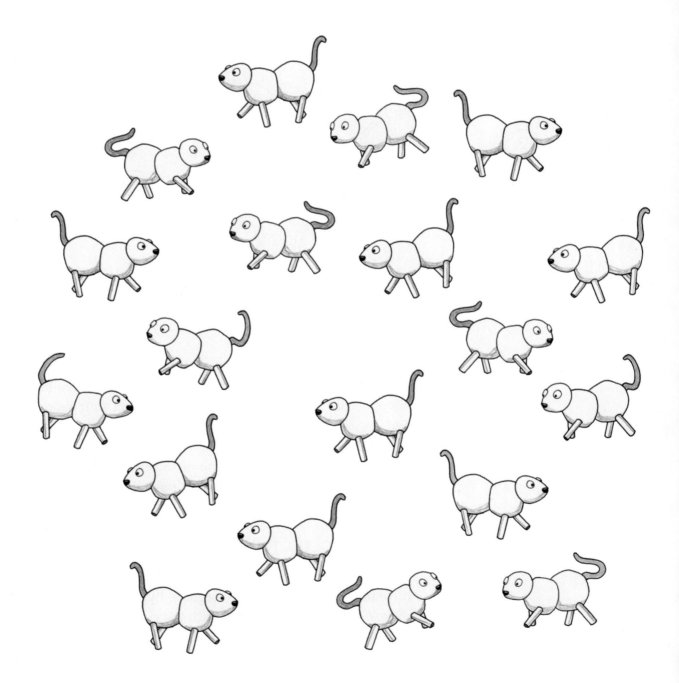

STOPPING TO THINK 3

Discuss with your group: Does the evidence so far from the second and third generations help you decide which hypothesis or hypotheses might be correct? Explain.

ANALYSIS

1. Based on the breeding results and your simulations, which hypothesis do you think best fits the evidence? Explain your answer.

2. **Reflection:** You have used models to investigate several scientific questions in *Science and Life Issues*. What are the trade-offs of using models to investigate the real world?

INVESTIGATION

An Austrian monk named Gregor Mendel studied the genetics of pea plants in the 1860s. Based on the results of his experiments with seed color and other pea traits, Mendel proposed a model for how organisms inherit traits from their parents.

You have explored a few hypotheses to explain the critter-breeding data. In this activity, you will investigate a hypothesis that is very similar to Mendel's hypothesis. Tossing coins is one way to model how genes are passed from parent to offspring. All of Skye and Poppy's offspring, including two named Ocean and Lucy, have blue tails. Why do some of Ocean and Lucy's offspring have orange tails? Find out by modeling the tail colors of the Generation Three offspring.

THE COIN-TOSSING MODEL

a. The outcome of a coin toss (heads or tails) represents the one version of a tail-color gene that is contained in the sex cell (sperm or egg) contributed by a parent critter. Heads represents the blue version and tails represents the orange version.

b. A future offspring critter receives a version of the tail-color gene from each of its two parents when fertilization occurs.

c. Each side of the coin represents one of the two versions of the tail-color gene carried by each Generation Two critter, such as Ocean and Lucy.

d. Blue tail color is **dominant** to orange tail color. This means that if a critter has at least one copy of the blue version of the gene, its tail is blue. A critter has an orange tail only if it has no blue versions of the tail-color gene.

CHALLENGE

How can tossing coins help you understand how organisms inherit genes from their parents?

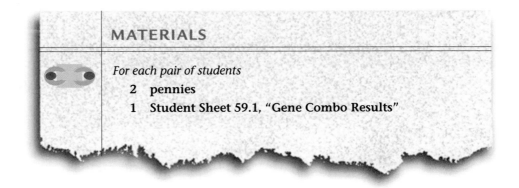

MATERIALS

For each pair of students
2 pennies
1 Student Sheet 59.1, "Gene Combo Results"

PROCEDURE

1. In your group of four students, divide into two pairs. Each pair should complete Steps 2–5.

2. Decide who will toss a penny to represent Ocean; the other person will toss a penny to represent Lucy. The outcome of each toss determines the tail-color gene each parent passes on to each critter pup.

3. Each person in your pair will toss a penny. For each toss, each partner should:

   • Hold a penny in cupped hands.

   • Shake it to the count of ten.

   • Allow it to drop from a height of about 20–40 cm (8–16 inches) onto the desk. Leave the pennies on the desk until you have completed Step 4.

4. Work with your partner to fill in the first row of Student Sheet 59.1, "Gene Combo Results," as shown below. Use these symbols to keep track of the genes:

   $\underline{T}$ = blue tail gene

   t = orange tail gene

$\succ$

(You could use any letter you like, but **T** and **t** can remind you of "tail color." To make your gene symbols easy to tell apart we suggest you always underline the uppercase letter of the pair.)

For example, if Ocean's coin toss results in heads (T), and Lucy's coin toss results in tails (t), your first entry will be:

Table 1: Gene Combo Results

Offspring	Ocean's Contribution (T or t?)	Lucy's Contribution (T or t?)	Offspring's Genes (TT, Tt, tT, or tt?)	Offspring's Tail Color (blue or orange?)
1	T	t	Tt	blue

Remember:
TT = *blue tail*
Tt, tT = *blue tail*
tt = *orange tail*

5. Repeat Steps 2–4 until you have filled in every row of the table.

6. Get together with the other pair in your group to prepare a simple table summarizing your results. Include the total number of times you got each gene combo (**TT**, **Tt**, **tT**, or **tt**) and the number of times you got each tail color (blue or orange).

7. Report the summary of your results to your teacher.

8. Add another row to your table to record the class data.

ANALYSIS

1. What is the ratio of blue-tailed to orange-tailed critter pups? Use the class data to answer this question:

   a. Divide the number of blue-tailed offspring by the number of orange-tailed offspring.

   $$ratio\ of\ tail\ colors = \frac{number\ of\ blue\text{-}tailed\ offspring}{number\ of\ orange\text{-}tailed\ offspring}$$

   b. Round this value to the nearest whole number. Then express it as a ratio by writing it like this:

   $$\underline{\hspace{3cm}} : 1$$
   (whole number)

c. Express this ratio as a pair of fractions, so that you can use them to complete the following sentence:

"About ____ of the offspring have blue tails, and about ____ of the offspring have orange tails."

d. Explain why the class obtained such a large ratio. For example, why isn't the ratio of blue to orange tails 1:1, that is, ½ blue and ½ orange?

2. You and your partner are about to toss two coins 100 times. Predict about how many times the outcome would be:

    a. heads-heads

    b. heads-tails

    c. tails-heads

    d. tails-tails

3. How sure are you that you will get exactly the results you predicted for Question 3? Explain your answer.

4. Look back at Activity 58, "Creature Features." Do the results of the coin-tossing model match the Generation Three critter data? Explain.

5. Try to write your own definition of the phrase *dominant trait* as it is used in genetics. **Hint:** Does it mean that every time any pair of critters mates, most of the offspring will have blue tails? Why or why not?

EXTENSION

For a larger sample size, combine your data with data gathered by fellow students from other locations. Go to the SALI page of the SEPUP website for instructions.

In Activity 59, "Gene Combo," you investigated a model for how inheritance works. This model is based on the work of Gregor Mendel. He discovered the behavior of genes 40 years before scientists learned where genes are located in the cell and almost 100 years before scientists discovered what the genes are made of!

CHALLENGE

Who was Gregor Mendel, and how did he discover the basis of heredity by breeding pea plants?

READING

Part One: Mendel's Life

In 1865 an Austrian monk named Gregor Mendel published his work on the behavior of genes. Mendel lived in a monastery that was devoted to teaching and science, as well as religious matters. Mendel had prepared to be a science teacher and had studied math, botany, and plant breeding for several years. The monastery had an experimental garden for agricultural research. Research then, as now, included breeding varieties of plants and animals in order to produce superior food and other

GREGOR MENDEL, 1866

products. Mendel became interested in the pea plants he was working on. He wondered how the colors, shapes, and heights of offspring pea plants were related to those of the parent plants. He began a careful study of how characteristics are inherited in pea plants. Mendel's discoveries depended upon research about breeding plants, careful experiments, and creative thinking. He understood that plant seeds are produced by sexual reproduction. He also understood the mathematics of probability.

His experiments were planned so carefully and recorded so thoroughly that anyone could repeat his procedure and confirm his findings. The patterns he discovered apply to all sexually reproducing organisms, from petunias to peas, from earthworms to humans. His discoveries have helped people answer questions ranging from how to produce disease-resistant food crops to explaining human diseases such as cystic fibrosis and sickle-cell disease.

STOPPING TO THINK 1

What personal qualities do you think Mendel must have had that helped him in his work?

Part Two: Mendel's Experiments with Pea Plants

Plants in general, and pea plants in particular, were excellent organisms for Mendel to study, for the following reasons:

- They grow into mature plants very quickly; in about 60–80 days their seeds grow and develop into mature plants.

- They produce numerous seeds rapidly.

- They have many observable characteristics that come in just two alternatives (such as purple or white flowers, yellow or green pods), with no blending of these traits (no lavender flowers, no yellowish green pods).

Table 1: Mendel's Results

	Flower Color	Seed Color	Seed Surface	Pod Color
Original Cross (Generation One)	purple x white	green x yellow	wrinkled x smooth	green x yellow
Generation Two Offspring	all purple	all yellow	all smooth	all green
Generation Three Offspring	705 : 224 (purple:white)	6,022 : 2,001 (yellow:green)	5,474 : 1,850 (smooth:wrinkled)	428 : 152 (green:yellow)

Mendel decided to breed extremely large numbers of pea plants and search for simple patterns in the offspring. The table above shows the results he obtained for four of the characteristics he studied.

Notice that Mendel did hundreds of crosses, and observed and counted the offsprings' traits for thousands of pea seeds, pods, and flowers. He then analyzed all these results and applied his knowledge of statistics and probability to infer the behavior of individual genes.

STOPPING TO THINK 2

a. What were the advantages for Mendel in using pea plants for his breeding investigations?

b. Why did Mendel perform so many crosses for the same characteristics?

As he looked at the data, Mendel noticed an interesting relationship. If he calculated the ratio of the two traits in the third generation, he obtained a ratio very near to 3:1.

For example, for seed color the ratio of yellow to green seeds can be calculated as

$$\frac{6{,}022 \; yellow}{2{,}001 \; green} = \frac{3.01 \; yellow}{1 \; green}$$

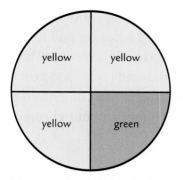

Figure 1: Seed Color Ratios

This is almost exactly a 3:1 ratio of yellow:green. This means that for every 1 green-seeded plant, there were almost exactly 3 yellow-seeded plants in the third generation.

You can also express this in terms of the fractions ¾ yellow and ¼ green, as shown in the pie graph in Figure 1.

Mendel concluded that not only did the green-seeded trait reappear in the third generation, but the probability of a third-generation plant having that trait was ¼—that is, about one green-seeded plant for every four plants produced overall.

He found the same ratio with other characteristics also. The 3:1 ratio he found for all of these different characteristics was the clue to how the parents' genes combine in their offspring. Based on his analysis, he proposed that:

- Each trait that appears in the second generation is the dominant version of the characteristic. The trait that is "hidden" is called **recessive**.

- Every plant has two copies of the gene for each characteristic. These two copies are called **alleles**. (The evidence suggested there was more than one copy of the gene for each characteristic in each plant, so the simplest assumption that would work was to have two alleles, or copies, per characteristic.)

- Each pea plant receives only one allele for each characteristic from each parent and ends up with two alleles of its own.

He then used these ideas to argue that the 3:1 ratio is exactly what is expected in the third generation for every plant characteristic. In the last two activities, you have explored this proposal, using disks and coins as models for genes.

STOPPING TO THINK 3

Explain how the model in Activity 59, "Gene Combo," works exactly like Mendel's explanation for his results with pea plants.

Mendel published his results in a paper that was mostly ignored at the time. Scientists and breeders failed to understand what he had done. Never before had someone used mathematics to tackle a complex biological problem. Mendel's work was rediscovered and understood only after scientists realized that genes must be located on the chromosomes found in the nucleus of the cell. You will learn about chromosomes in Activity 63, "Show Me the Genes!"

ANALYSIS

1. Based on Mendel's results, what trait for each pea characteristic is dominant? Make a table of the dominant and recessive alternatives for each characteristic in Table 1. Add an extra column to your table; you will use it to record your answers to Question 2a.

2. **a.** Calculate to the hundredths place the ratio of dominant to recessive for each characteristic in the third generation. Record the ratio for each characteristic in the table you prepared for Question 1.

   b. Why are the ratios not exactly 3:1?

3. Look at Figure 1, which shows the ratio of green-seeded and yellow-seeded offspring. Explain why a 1:3 ratio of green-seeded plants to yellow-seeded plants is the same as a fraction of ¼ green-seeded plants.

4. Mendel performed his experiments on more characteristics than the four shown in Figure 1. Why was it important for him to look at more than one characteristic?

5. **Reflection:** People often think of mathematics as important to physics and chemistry, but not to life science (biology). What is your opinion?

In previous activities you have learned that parents contribute half of their genes to each offspring. You have also used coin tosses to model the way genes are passed on to the offspring of two parents. In this activity you will learn how a table, called a Punnett square, can help you predict the results of a breeding experiment.

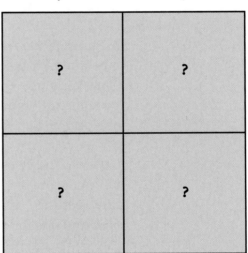

CHALLENGE

How can Punnett squares help you predict patterns of inheritance?

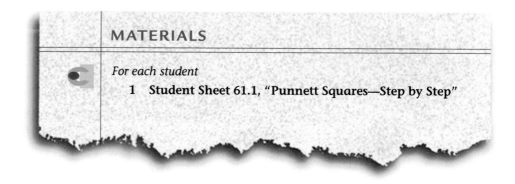

MATERIALS

For each student
1   Student Sheet 61.1, "Punnett Squares—Step by Step"

PROCEDURE

Read "How to Use Punnett Squares," and then complete the Punnett squares on Student Sheet 61.1.

HOW TO USE PUNNETT SQUARES

A **Punnett square** is a diagram you can use to show how likely each outcome of a breeding experiment is. It is used when each parent's genes for a trait are known. By filling in the squares, you can find the possible combinations of genes in the offspring of the two parents. You can also predict the chances that each kind of offspring will occur.

Consider the cross between Skye and Poppy as an example. The two tail colors are blue and orange. As you modeled in Activity 58, "Creature Features," there are two versions of the tail-color gene, one for blue and one for orange. These two versions are called alleles. As you saw in Activity 59, "Gene Combo," the blue allele is written as uppercase **T** and the orange allele as lowercase **t**. This is because we know that blue tail color is dominant to orange. (You might also use **B** for blue and **b** for orange, since blue is the dominant trait. But you need to use the same letter, uppercase and lowercase, for the two alleles of any one gene. To avoid confusion, always remember to underline the uppercase letter.)

Because Skye is from an island where there are no orange-tailed critters, we can assume he has only blue tail-color alleles. So, his alleles for tail color are **TT**. Because Poppy is from an island where there are no blue-tailed critters, we can assume she has only orange tail-color alleles. So, her alleles for tail color are **tt**. An organism that has only one kind of allele for a characteristic is called **homozygous**. Skye is homozygous for the blue tail-color trait, while Poppy is homozygous for the orange tail-color trait.

STEP 1: STARTING A PUNNETT SQUARE

Write the possible alleles donated by each parent along the top of the table and left side of the table—it doesn't matter which parent you use for each position. In the table below, Skye's alleles are placed along the top and Poppy's at the left.

Each **T** along the top represents an allele in the sperm cell produced by Skye. Each **t** on the left represents an allele in the egg cell from Poppy.

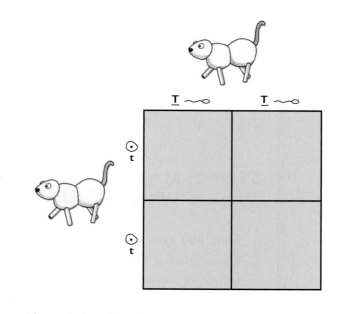

Figure 1: Starting a Punnett Square
Remember: an underlined uppercase letter is used for the allele for the dominant trait. A lowercase letter is used for the allele for the recessive trait.

STEP 2: COMPLETING A PUNNETT SQUARE

Complete each box of the table by combining one allele from the top and one allele from the left, as shown below.

When you combine one allele from each parent into a box, you are representing a sperm cell fertilizing an egg.

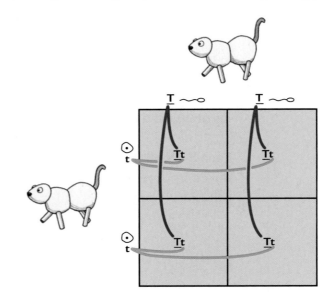

Figure 2: Completing a Punnett Square

STEP 3: MAKING CONCLUSIONS USING A PUNNETT SQUARE

Now you can use the Punnett square to make some conclusions. All the offspring of Skye and Poppy will have one blue tail-color allele and one orange tail-color allele. An organism that has alleles for two different traits is called **heterozygous**. Because blue tail color is dominant over orange, all offspring will have blue tails (as was found in the breeding experiment between Skye and Poppy).

ANALYSIS

1. Compare the results of your Punnett square for Problem 1 on Student Sheet 61.1 with the results of the Ocean/Lucy cross in Activity 59, "Gene Combo." Why are they similar?

2. Refer to the table of Mendel's results in Activity 60, "Mendel, First Geneticist," on page D-36.

   a. What are the traits for pea flower color? Suggest letters you might use to represent the alleles for flower color.

   b. What are the traits for seed surface? Suggest letters you might use to represent the alleles for seed surface.

3. Review your results on Student Sheet 61.1. Why is it impossible for offspring to show the recessive trait if one parent is homozygous for the dominant trait?

4. A scientist has some purple-flowered pea plants. She wants to find out if the pea plants are homozygous for the purple flower color.

   a. What cross will be best to find out if the purple-flowered peas are homozygous?

   b. Use Punnett squares to show what will happen if the plants are crossed with white-flowered plants and

   i. the purple-flowered plants do not have an allele for the white trait.

   ii. the purple-flowered plants do have an allele for the white trait.

LABORATORY

I n this activity, you will obtain data from the seeds you germinated in Activity 55, "Plants Have Genes, Too!" and investigate whether your own data fit Mendel's model for inheritance.

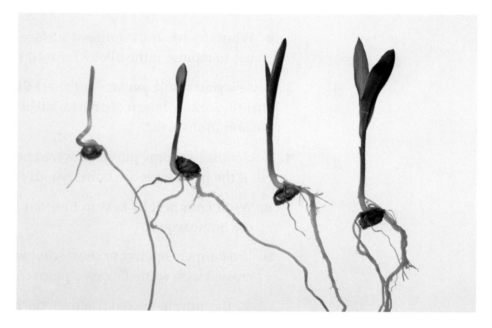

CHALLENGE ⟶

Do your results fit with Mendel's model for inheritance?

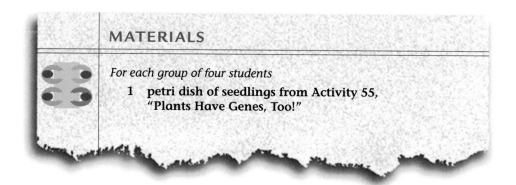

MATERIALS

For each group of four students
1   petri dish of seedlings from Activity 55,
     "Plants Have Genes, Too!"

PROCEDURE

1. Review the prediction(s) you made when you began Activity 55. If you would like to change your prediction, record your revised prediction in your science notebook.

2. Get your dish of sprouted seedlings.

3. With the lid off, examine each seedling plant carefully. With your partner, count the number of green seedlings and the number of yellow seedlings.

4. Prepare a data table in your science notebook to summarize your results. Add an extra line for the class total results.

5. Report your results to your teacher, as directed.

6. Your teacher will display the class's total data. Record the data in your data table.

ANALYSIS

1. Look back in your science notebook for the prediction you made in Activity 55. Was your prediction correct? Explain.

2. Compare the class's results for seedling color to Mendel's results for various pea plant traits. Why are they similar? What do they suggest about the inheritance of the pale yellow and green *Nicotiana* traits?

3. Do each group's results fit Mendel's model? Explain.

4. When you first set out these seeds to germinate, you were told that they were all the offspring of two green parent plants. You were also told that each of the green parents had one green parent and one yellow parent.

a. Based on the class's results, what can you conclude about the color alleles of each of the green parents of your seedlings?

b. How is this breeding cross similar to the one you modeled in Activity 59, "Gene Combo"? Explain.

5. Construct a Punnett square to show what will happen if one of the green parent-generation *Nicotiana* plants is crossed with a pale yellow plant. Explain the results.

EXTENSION

For a larger sample size, combine your data with data gathered by fellow students from other locations. Go to the SALI page of the SEPUP website for instructions.

READING

Even before he began his experiments on pea plants in the 1860s, Mendel knew that the genes had to be in the male and female sex cells—the sperm (or pollen) and the egg. However, no one knew *where in the cell* the genes were located until fifty years later. An individual gene is too small to be seen with a light microscope. So how did the microscope enable scientists to figure out where in the cell the genes are located? Scientists studying dividing cells, like the one shown in the photo below, provided evidence to support Mendel's ideas.

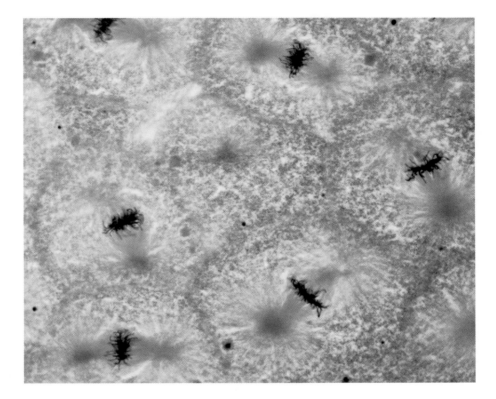

CHALLENGE

What role do chromosomes play in the inheritance of genes?

READING

Part One: Chromosomes and Cell Division

Every organism must make new cells. Single-celled organisms, such as bacteria, yeast, paramecia, and amoebas, use cell division to reproduce asexually. In multicellular organisms, cell division (Figure 1) is necessary for the organism to grow to adulthood and to replace injured and worn out cells. When you get a cut and lose some blood, additional new blood cells and new skin cells are produced from the division of cells in your body.

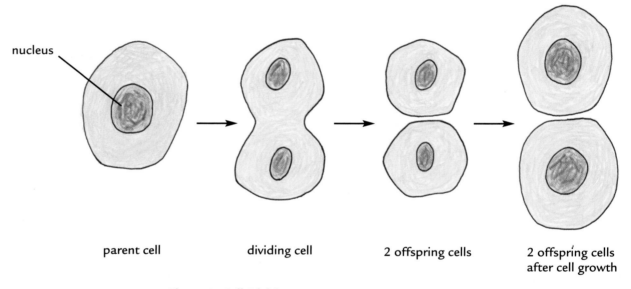

nucleus

parent cell   dividing cell   2 offspring cells   2 offspring cells
                                                                        after cell growth

Figure 1: Cell Division

STOPPING TO THINK 1

How is the function of cell division in single-celled organisms different from cell division in multicellular organisms?

In the early 1900s, scientists studying cells in rapidly growing parts of plants made an interesting observation. They saw that just before cell division, the membrane around the nucleus was no longer visible and little dark structures, which they called **chromosomes**, appeared. When the cells split apart, the chromosomes were divided evenly between the two new cells (see Figure 2).

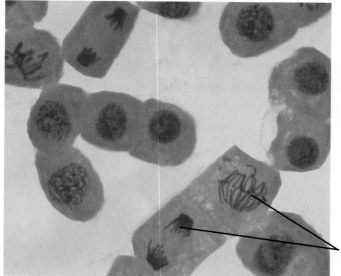

chromosomes

Figure 2: Chromosomes in Dividing Plant Cells

When a cell is not dividing, the chromosomes are long, fine strands, like very thin spaghetti, packed into the nucleus of the cell. Before the cell divides, it makes copies of all its chromosomes so that its two offspring cells can each get a complete set. Then the chromosomes become coiled, which makes them visible when observed under a microscope. Finally, the cell divides, as shown in Figure 3.

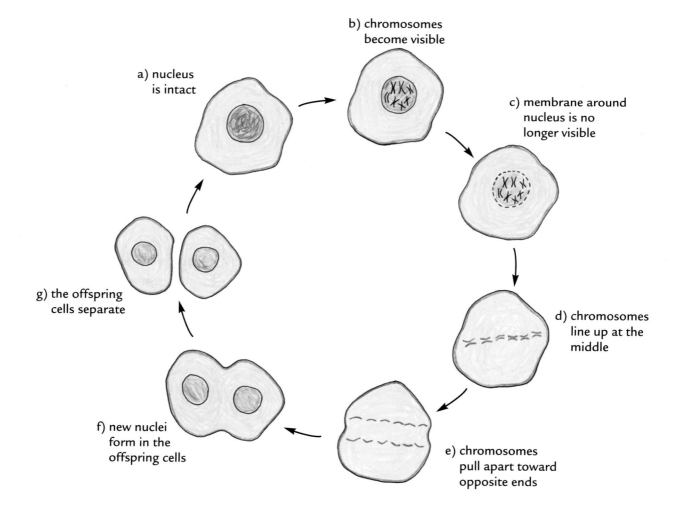

a) nucleus is intact

b) chromosomes become visible

c) membrane around nucleus is no longer visible

d) chromosomes line up at the middle

e) chromosomes pull apart toward opposite ends

f) new nuclei form in the offspring cells

g) the offspring cells separate

Figure 3: Cell Division in More Detail

Each cell in a human body contains 46 chromosomes. When a human cell divides, the two cells that result each contain 46 chromosomes. How can 46 chromosomes become two sets of 46? It's not magic: each crisscrossed chromosome is two identical copies that are attached to each other. As you can see in Figure 4, each doubled chromosome then splits during division to become two identical, but now separate, chromosomes.

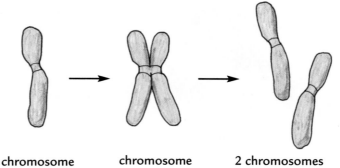

chromosome   chromosome   2 chromosomes
                          doubles

Figure 4: A Chromosome During Cell Division

STOPPING TO THINK 2

What would happen to the number of chromosomes in each cell if copies of them were not made before cell division?

Part Two: Chromosomes and Sexual Reproduction

The 46 chromosomes in a human cell can be sorted into 23 matching pairs, as shown in Figure 5. Each chromosome looks identical to its partner, with one exception: pair number 23, which are also called the sex chromosomes. Female humans have two X-chromosomes, while males have one X-chromosome and one Y-chromosome.

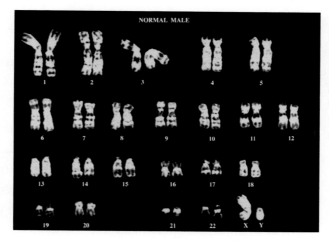

Figure 5: The 46 Chromosomes in Every Cell of a Male Human
These chromosomes were photographed in a flattened cell; the images of each chromosome were cut out and sorted by size.

Soon after the 23 pairs of chromosomes were observed, scientists declared that the 23 pairs of chromosomes behaved just like the genes in Mendel's model. What did they mean by that?

Sex cells (sperm and egg) are formed by a special kind of cell division, in which each cell receives copies of exactly half of the chromosomes. In humans, the body's cells contain 46 chromosomes, but the egg and sperm contain only 23 chromosomes. When a sperm fertilizes an egg cell to form the first cell of a new organism, the new cell has 46 chromosomes, as illustrated in Figure 6. Half come from the mother and half come from the father.

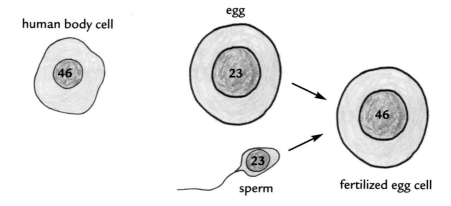

Figure 6: Sex Cells and Fertilization

....

STOPPING TO THINK 3

Why must the number of chromosomes in the sperm and egg be half the number of chromosomes in the other cells of an organism?

....

When sex cells are produced, exactly one member of each chromosome pair moves into each offspring cell. Thus, when an egg and a sperm cell come together in fertilization, the new cell has 23 complete pairs of chromosomes, or 46 total chromosomes. Think of the 23 pairs of chromosomes as 23 pairs of socks. An egg cell has one of each kind of sock, and a sperm cell has the matching sock for each of the egg's socks. Only when they come together, when the sperm fertilizes the egg, does the resulting cell contain 23 complete pairs of socks. One member of each pair has come from the female parent, and the other from the male parent. The chromosomes of each pair carry genes for the same characteristics (see Figure 7), but the two alleles of any one gene can be different, as you've learned before.

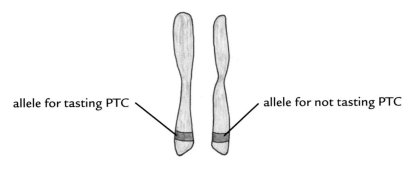

allele for tasting PTC   allele for not tasting PTC

Figure 7: Alleles in Humans for the Ability to Taste PTC
A matching pair of chromosomes contains two different alleles for PTC tasting, since this PTC-tasting person is heterozygous. The two alleles of the gene might differ in a single bit of genetic information.

By comparing the microscope evidence to the work of Gregor Mendel, scientists realized that the chromosomes must carry the genes. Once scientists understood the location of the genes and the way they were passed to offspring, they realized the importance of Mendel's work.

STOPPING TO THINK 4

Consider two children with the same two parents. Would you expect them to have the same sets of chromosomes? Explain why or why not.

Part Three: So Many Genes . . .

Human cells contain approximately 30,000 pairs of different genes. However, the human nucleus contains only 23 pairs of chromosomes. Each gene is a small portion of a chromosome. Only by careful study can scientists determine which gene (or group of genes) is responsible for a specific trait, such as eye color or blood type in humans or seed color in pea plants.

Furthermore, the more complex the trait being studied, the greater the number of different genes which contribute to it. Even diseases such as cancer are proving to result from the combined effects of many genes.

The genes are part of a long molecule called DNA, which stands for deoxyribonucleic acid. Each chromosome contains a long DNA molecule. Sometimes, before a cell divides, a mistake is made in a gene when the DNA is copied to make a new set of chromosomes. These mistakes are called **mutations**. If the mutation occurs during the formation of a sex cell, an offspring that results from that sex cell will be affected by the mutation. Genes give instructions to the body. Even though some mutations don't make much difference and some are even helpful, most mutations are harmful.

A mutation in a gene is like a change in a word in a sentence. For example, consider the sentence, "I hear that noise." The four letters in the word "hear" communicate a meaning. What if we change one of the letters? The results might include

I heer that noise.

I fear that noise.

The first change makes the word look a little funny, but it still sounds the same, and the meaning of the sentence would be unchanged if you heard it spoken aloud. But the second change completely changes

➢

the meaning of the sentence. Like this change in the sentence, some mutations change the information provided by the genes in which they occur.

....

STOPPING TO THINK 5

How exactly does a mutation change the form of an organism? When do such mutations occur?

....

ANALYSIS

1. Draw a flow diagram (a series of pictures) such as the one below that shows the locations and relative sizes of DNA, genes, chromosomes, and cells in a human body. Write a paragraph to explain your diagram.

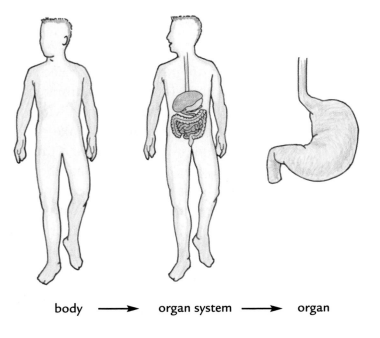

body   ⟶   organ system   ⟶   organ

LABORATORY

Genes help determine everything about you. But how big a role do they play? Magazine and newspaper articles often describe the impact of genes and the environment on the development of human traits. Sometimes these articles refer to this as the "Heredity vs. Environment" or "Nature vs. Nurture" debate. In fact, this is not a debate that one side will "win." Your genes and your environment *both* make you a unique person. Researchers would like to know more about how genes and the environment affect human traits.

Investigations of other organisms provide examples of how genes and the environment affect traits. The shapes of trees, the colors of

hydrangea flowers, and the dark coloring at the tips of a Siamese cat's ears are all traits determined by both heredity and the environment. In this activity, you will use *Nicotiana* seeds to investigate how genes and the environment affect plant growth and development.

The cooler temperature at the tips of the cat's ears, face, paws, and tail is necessary for the dark color to develop, even though the gene for dark color is in every cell of the cat.

These hydrangea flowers can be either pink or blue, depending on soil conditions.

CHALLENGE

How does the environment affect the inherited green color trait in *Nicotiana*?

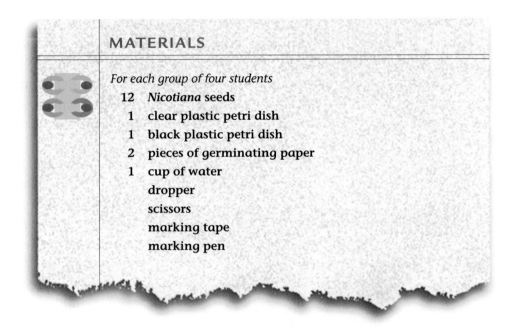

MATERIALS

For each group of four students

12	*Nicotiana* seeds
1	clear plastic petri dish
1	black plastic petri dish
2	pieces of germinating paper
1	cup of water
	dropper
	scissors
	marking tape
	marking pen

PROCEDURE

Part One: Planting the Seeds

1. Review the Materials list. With your group, design an experiment to test the effect of light on the inherited green color trait in *Nicotiana*.

   When designing your experiment, think about the following questions:

   - What is the purpose of your experiment?

   - What variable are you testing?

   - What is your hypothesis?

   - What variables will you keep the same?

   - What is your control?

   - How many trials will you conduct?

- Will you collect qualitative and/or quantitative data? How will these data help you form a conclusion?

- How will you record these data?

2. Record your hypothesis and your planned experimental procedure in your science notebook.

3. Make a data table that has space for all the data you need to record. You will fill it in during your experiment.

4. Obtain your teacher's approval of your experiment.

5. Conduct your experiment and check your seedlings every day. As with your earlier experiment, wait until the seedlings are old enough for you to be able to see which color trait they have developed.

Part Two: Analyzing Results

6. Collect your seedlings.

7. Observe your seedlings carefully to see if there are any differences. Record your observations in the data table you prepared when you set up your experiment.

8. Report your results to your teacher, as directed.

9. After analyzing your results, respond to the questions below.

ANALYSIS

1. Was your hypothesis correct? Explain.

2. What effect did heredity have in determining the color of the seedlings?

3. What effect did the environment have in determining the color of the seedlings?

4. Can heredity alone ensure that an organism will grow well and be healthy? Explain.

5. Can the environment alone ensure that an organism will grow well and be healthy? Explain.

6. **Reflection:** What role do you think genes and the environment play in human development and health? Explain your thinking and give some examples.

➤

EXTENSION

For a larger sample size, combine your data with data gathered by fellow students from other locations. Go to the SALI page of the SEPUP website for instructions.

INVESTIGATION

When you considered Skye and Poppy's breeding, you focused on their tail colors. The tail-color inheritance followed the same pattern described by Mendel in his pea plant inheritance experiments. One gene, for which there are two alleles, determines each characteristic. Each characteristic has two versions, or traits. For the characteristic of tail color, the traits are blue and orange, and the alleles of the tail-color gene are referred to as **T** and **t**. One trait is completely dominant over the other, recessive, trait. In this case, blue tail color (**T**) is dominant over orange (**t**).

In fact, Skye and Poppy have a variety of traits. Some of them follow the pattern of inheritance that was described by Mendel. Others are inherited in a slightly different pattern. In this activity, you will look at some other traits and investigate how they are inherited.

Skye

Poppy

Lucy

Ocean

CHALLENGE

What are some patterns of inheritance other than the one discovered by Mendel?

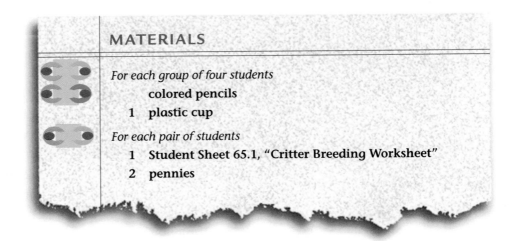

MATERIALS

For each group of four students
  colored pencils
1 plastic cup

For each pair of students
1 Student Sheet 65.1, "Critter Breeding Worksheet"
2 pennies

THE MODEL

The table below shows Skye's and Poppy's traits. It also shows the traits of all their offspring. In this activity, you will look at more traits in the Generation Three offspring, which are produced when Generation Two offspring mate with each other. Lucy, a female, and Ocean, a male, are the Generation Two critters who will mate.

Table 1: Generation One and Generation Two Traits			
Characteristic	**Skye**	**Poppy**	**100 Offspring (such as Lucy and Ocean)**
Body segments (number)	2	3	3
Leg color	blue	red	blue
Eyes (number)	2	3	2
Nose length	short	long	medium
Tail color	blue	orange	blue
Tail style	straight	curly	48 curly, 52 straight
Antennas (number)	1	2	2
Spikes (color and number)	1 short blue	2 long green	1 short blue + 2 long green
Sex	male	female	53 female, 47 male

PROCEDURE

1. Work in pairs. Place Student Sheet 65.1, "Critter Breeding Worksheet," between you and your partner. The person sitting on the left side will toss a penny for Ocean, while the person on the right will toss a penny for Lucy.

Table 2: Critter Code		
Characteristic	**Alleles**	**Trait**
Body segments (number)	**BB**	3
	Bb	3
	bb	2
Leg color	**LL**	blue
	Ll	blue
	ll	red
Eyes (number)	**EE**	2
	Ee	2
	ee	3
Nose length	**NN**	long
	Nn	medium
	nn	short
Tail color	**TT**	blue
	Tt	blue
	tt	orange
Tail style	**SS**	curly
	Ss	*curly or straight
	ss	straight
Antennas (number)	**AA**	2
	Aa	2
	aa	1
Spikes (color and number)	**GG**	1 short blue
	HH	2 long green
	GH	1 short blue + 2 long green

To find out if an Ss critter's tail is curly or straight, toss a coin. If it shows heads, the critter's diet contains "crittric" acid, and it develops a curly tail. If the coin shows tails, the critter's diet does not contain "crittric" acid, and it develops a straight tail.

2. For each toss, each partner should:

   • Hold a penny in cupped hands.

   • Shake it to the count of ten.

   • Allow it to drop from a height of about 20–40 cm (8–16 inches) onto the desk.

3. The partner on the left tosses a penny to determine which allele for number of body segments Ocean gives to his offspring. If the penny shows heads, write _B_ in the column titled "From Ocean" on Student Sheet 65.1. If the penny shows tails, write _b_. The other partner tosses a penny to determine the allele which Lucy gives. Write the letter for that allele in the column titled "From Lucy."

4. Determine the offspring's trait for number of body segments. Look at the alleles you wrote under "From Ocean" and "From Lucy." Compare these alleles with the Critter Code in Table 2 (or with the information in the first column of Student Sheet 65.1). Then write the appropriate trait in the next column. For example, if you wrote B͟b for the alleles, the trait is "3 segments."

5. Continue tossing coins and filling in Student Sheet 65.1 until you have completed the table. Use the Critter Code to determine the trait for each characteristic, based on the allele combination in the offspring. Note the special instructions for tail style.

➢

6. Find out if your critter is male or female by determining its sex chromosomes as follows:

a. Ocean is an XY male. The partner representing Ocean tosses a penny. If it shows heads, Ocean donates an X chromosome to the offspring. If the penny shows tails, Ocean donates a Y chromosome to the offspring.

b. Lucy is an XX female. The partner representing Lucy does not need to toss a penny. Lucy can donate only an X chromosome to the offspring.

Write the sex (male or female) of the offspring in the appropriate space.

7. Use the following materials provided by your teacher to make your critter.

Table 3: Critter Parts	
Body segments	Large foam balls connected by toothpicks
Heads	Small foam balls
Legs	Pieces of red or blue drinking straws
Eyes	Blue thumbtacks
Nose	Brass fastener, adjust length
Tail	Blue or orange pipe cleaner
Antennas	Yellow paper clip
Spikes	Pieces of blue or green drinking straws

8. Draw your critter and color in the body parts.

ANALYSIS

1. Look at the other critters made by your classmates. They are all siblings (brothers and sisters). What are their similarities and differences?

2. Which characteristics show a simple dominant/recessive pattern like tail color? List them in a table and indicate which version is dominant and which is recessive for each trait.

   Hint: Look at Table 1 to see which traits have this pattern.

   Some traits do not show a simple dominant vs. recessive pattern. Look at Table 1 to help you answer Questions 3–5.

3. For which characteristic do some offspring have traits in between Skye's and Poppy's traits? Explain. (For example, in some plants, a cross between a red- and white-flowered plant will give pink-flowered offspring. This is called incomplete dominance.)

4. For which characteristic do some offspring have both Skye's and Poppy's traits? Explain. (For example, in humans, a person with type A blood and a person with type B blood can have a child with type AB blood. This is called co-dominance, as both traits appear in the offspring.)

5. Which critter trait is affected by an environmental factor, such as light, temperature, or diet? Explain.

6. Consider the pattern for sex determination.

   a. How is a critter's sex determined?

   b. Whose genetic contribution—Ocean's or Lucy's—determines the sex of the offspring?

7. Who does your critter most look like—Skye, Poppy, Ocean, or Lucy? On which traits did you base your choice?

8. Draw a critter with all recessive traits. Assume the recessive trait for spikes is no spikes.

As you now know, genes are inherited and affect the characteristics of an organism. By growing *Nicotiana* seedlings, you've seen how a trait is inherited. You have also seen how Punnett squares can help make predictions about inherited traits in large numbers of offspring.

Studying human inheritance is more difficult. Scientists cannot perform breeding experiments on people. They must use other approaches when studying human genetics. Family histories, such as this one, provide one way to gather evidence about inherited traits in humans.

A Partial Pedigree of Hemophilia in the Royal Families of Europe

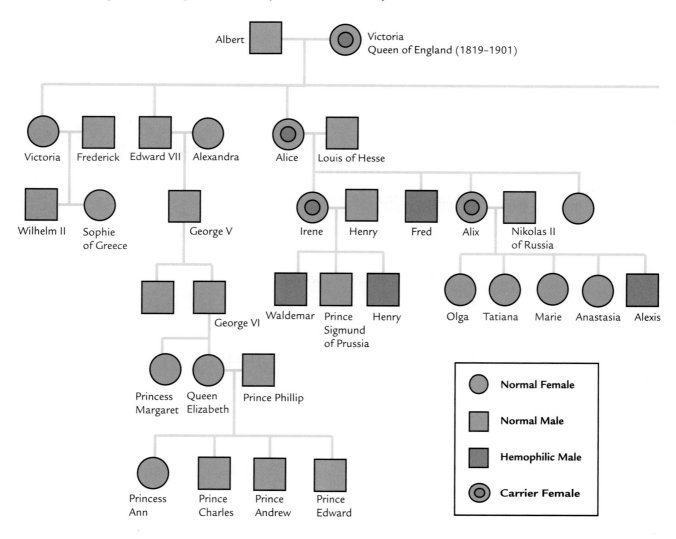

CHALLENGE

How can you use pedigrees to study human traits?

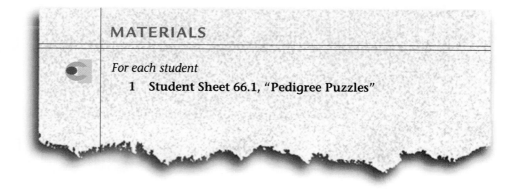

MATERIALS

For each student
1   Student Sheet 66.1, "Pedigree Puzzles"

PROCEDURE

Work with your group to read and discuss pedigrees.

What is a Pedigree?

One approach to studying genes in humans and other organisms is to collect data on a single trait within a family. These data can be used to construct a family tree used for genetic analysis. Such a tree is called a **pedigree**. Researchers use a pedigree to look for patterns of inheritance from one generation to the next. These patterns can provide clues to the way the trait is passed from parents to their offspring (or children).

Figure 1 on the next page shows the pedigree for tail color in the breeding family of Skye and Poppy. Squares are used to represent males and circles are used to represent females. The zoo breeders were concerned that the orange tail color trait might be lost when it did not appear in the second generation, but it reappeared in the third generation.

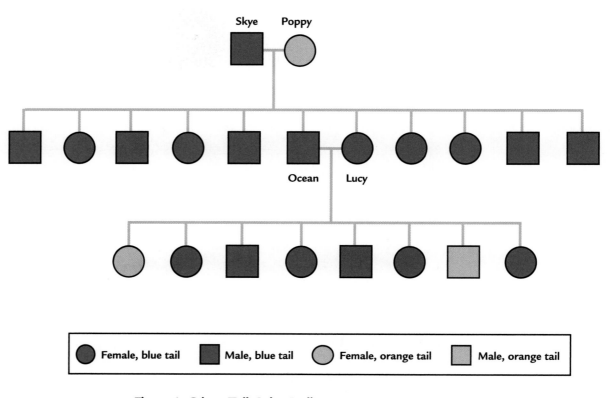

Figure 1: Critter Tail-Color Pedigree
This pedigree examines the tail-color trait in the family of critters bred in the zoo.

You have used Punnett squares and your knowledge of which trait is dominant to predict what fraction of the offspring of a particular pair of parents are likely to have each trait. But how do scientists find out which trait is dominant when they can't do breeding experiments? They analyze data provided by pedigrees.

STOPPING TO THINK 1

a. Look carefully at Figure 1. Explain how the information in the pedigree tells you whether orange tail color is dominant or recessive.

b. You have used the symbols **T** and **t** for the alleles of the critter tail-color gene. On Student Sheet 66.1, "Pedigree Puzzles," label each critter in the diagram with the gene combination you think it must have.

c. Why don't you know whether each blue-tailed critter in Generation Three is homozygous or heterozygous?

Learning from Data on Human Conditions

Most human traits, such as height, for example, are the results of interactions between many genes and environmental factors. But some hereditary diseases in humans, such as the Marfan syndrome (see Activity 56, "Joe's Dilemma"), are caused by a single gene. Pedigrees from several generations of a family enable scientists to figure out whether such a condition is dominant or recessive. Once doctors know this, they can predict how likely it is that a child of particular parents will have the condition.

For example, individuals with a condition called PKU, or phenylketonuria (feh-null-key-tun-YUR-ee-uh), cannot break down protein normally. This leads to the build-up of a chemical that causes mental retardation. If PKU is diagnosed shortly after birth, the child can be given a special diet. Children given this special diet for at least the first 10 years of life do not develop the symptoms of the condition. In most of the United States, newborns are routinely tested for PKU within a few days after birth.

Analysis of pedigrees indicates that PKU almost always appears in children of people who do *not* have the condition. Figure 2 shows a family in which two grandchildren inherited PKU and five did not. These numbers vary from family to family.

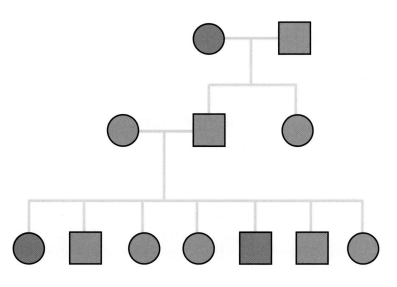

Figure 2: Family with PKU
Affected individuals are shown in red.

Is PKU likely to be a dominant or a recessive trait? How was it inherited by the individuals in the pedigree in Figure 2? On Student Sheet 66.1, label each individual with the allele combination(s) he or she might have.

Hint: Remember that if the condition is dominant, an affected individual could be homozygous or heterozygous. But if it's recessive, an affected individual must be homozygous for the trait.

In the study of genetic diseases, a person who is heterozygous for a recessive genetic condition is called a **carrier**. Such a person does not have the condition, but can pass on an allele for it to his or her children. The recessive allele is hidden, or masked—until it shows up in a homozygous individual who has the condition. A person who has a recessive condition is not called a carrier.

Note: In the Micro-Life unit of *Science and Life Issues,* you learned a very different use of the term *carrier.* When describing infectious diseases, a carrier is someone who does not show symptoms of an infectious disease but can infect other people with the microbe responsible for that disease.

STOPPING TO THINK 3

Why is it impossible for an individual to inherit a recessive condition if only one parent is a carrier for that condition?

Are All Hereditary Conditions Recessive?

PKU is a recessive trait; both parents must be carriers for a person to inherit it. Is this true for all other hereditary diseases, such as the Marfan syndrome? Consider polydactyly (paw-lee-DAK-tul-ee), which causes individuals to have an extra finger or toe on each hand or foot. This is not a dangerous condition, but it does run in families. Figure 3 shows a typical pedigree of a family with polydactyly. One grandchild in this family has polydactyly, but her three siblings do not. These numbers vary from family to family.

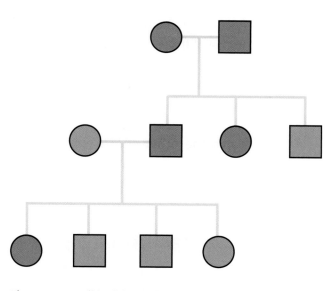

Figure 3: Family with Polydactyly
Affected individuals are shown in red.

STOPPING TO THINK 4

Is polydactyly likely to be a dominant or a recessive trait? How did the individuals in the pedigree in Figure 3 inherit it? On Student Sheet 66.1, label each individual with the allele combination(s) he or she might have.

Hint: Remember that if the condition is dominant, an affected individual could be homozygous or heterozygous. But if it's recessive, an affected individual must be homozygous for the trait.

Based on such pedigrees, scientists have concluded that polydactyly is a dominant trait. The condition does not skip a generation the way recessive traits can (see Figure 2).

The Genes for ABO Blood Groups

In certain plants, when purebred red flowers are crossed with purebred white flowers, the offspring are not red or white, but pink. These pink heterozygous flowers show both alleles they possess—the overall effect is an intermediate appearance. Red and white flowers are not dominant or recessive traits, but show **incomplete dominance**.

Recall the chemical PTC, which you tasted (or did not taste!) in Activity 54, "Investigating Human Traits." The ability to taste PTC is a dominant trait. However, there can be some incomplete dominance as well: heterozygous individuals may taste PTC less strongly than people with two copies of the tasting allele.

For some other characteristics, two traits both appear at the same time. Since both traits can be observed distinctly in a heterozygous individual, this is known as **co-dominance** (equal dominance).

STOPPING TO THINK 5

Look back at Activity 65, "Breeding Critters—More Traits." Which characteristic modeled incomplete dominance? Which characteristic modeled co-dominance?

In Activity 46, "Disease Fighters," you learned about the ABO blood groups. You saw that people with some blood types have an immune reaction to blood of certain other types; they cannot be given transfusions of these incompatible blood types. In Activity 68, "Searching for the Lost Children," you will see how blood typing can help solve real-life problems. There are *four* different blood types, A, B, O, and AB. How are blood groups inherited?

After investigating pedigrees from many families, scientists obtained the results listed in Table 1.

Table 1: ABO Blood Types		
Parents' Blood Types		**Children's Possible Blood Types**
O	O	O
A	O	A or O
B	O	B or O
A	A	A or O
B	B	B or O
A	B	AB, A, B, or O

Based on these results, figure out the possible allele pairs for each of the four blood types. Use Student Sheet 66.1 (page 2) to record your work.

Blood Type	Possible Allele Pairs
O	_____
A	_____ or _____
B	_____ or _____
AB	_____

STOPPING TO THINK 6

Which two blood types are co-dominant? Which blood type is recessive?

ANALYSIS

1. The following pedigrees represent the blood types in four unrelated families. In each case, the parents have Type A and Type B blood.

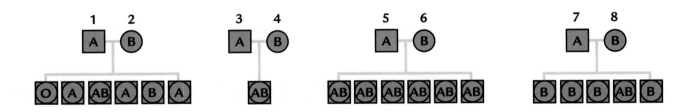

   a. Which of the eight parents are definitely heterozygous for the Type O allele? Explain.

   b. Which of the eight parents are probably not heterozygous for the Type O allele? Explain.

   c. Can you be certain that the parents you named in response to Question 1b do not have a Type O allele? Explain.

2. The pedigree shown below represents a genetic condition. Use the information it provides to answer the questions below. Use Student Sheet 66.1 (page 3) to try out allele combinations for related individuals.

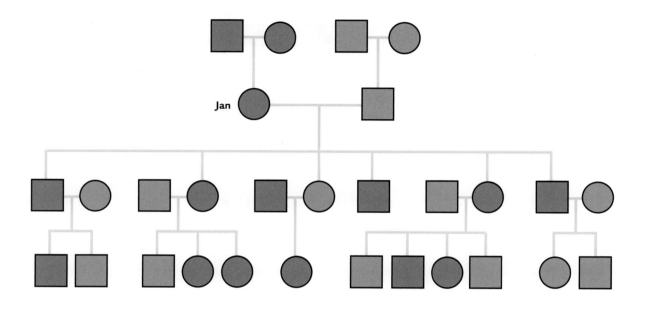

a. Is the condition most likely a dominant or a recessive trait? Explain your reasoning.

b. Is Jan most likely to be homozygous dominant, heterozygous, or homozygous recessive?

3. The pedigree shown below represents another genetic condition.

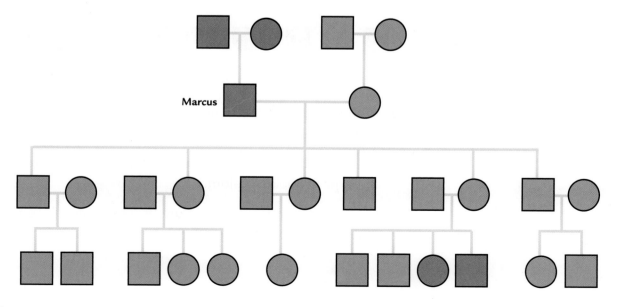

a. Is the condition most likely a dominant or a recessive trait? Explain your reasoning.

b. Is Marcus most likely to be homozygous dominant, heterozygous, or homozygous recessive?

4. The pedigree shown below represents a third genetic condition.

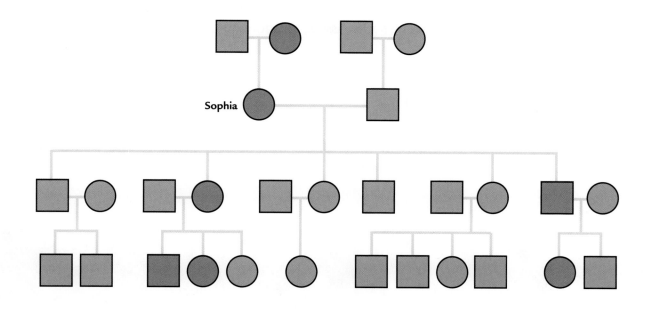

a. Is the condition most likely a dominant or a recessive trait? Explain your reasoning.

b. Is Sophia most likely to be homozygous dominant, heterozygous, or homozygous recessive?

5. The term *carrier* is used very differently in genetics than in the study of diseases.

a. What is being "carried" by a genetic carrier? What is being "carried" by a disease carrier?

b. How does transmission occur for genetic conditions? How does transmission occur for infectious diseases?

TALKING IT OVER

New tests are being developed for genetic conditions as scientists learn more about the genes that cause them. These tests will help people plan their lives, as well as lead to actions that help prevent some conditions from having serious effects. But these new tests also raise issues for individuals and for society.

CHALLENGE

How do individuals and society react to the issues raised by genetic testing?

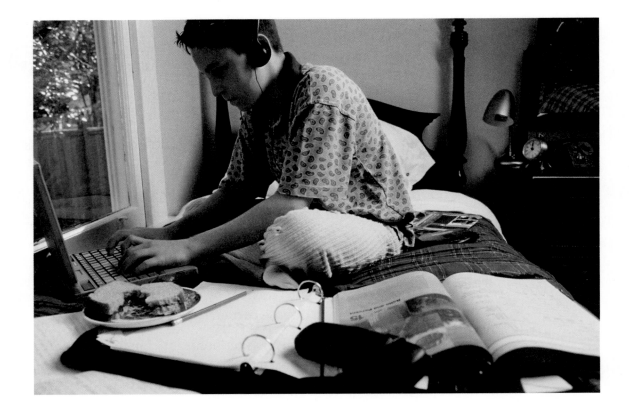

READING

to: meganR@talk.com

from: joeF@email.com

Megan—

I've learned more from the doctor and genetic counselor about this condition I might have. The Marfan syndrome is dominant—it's caused by a single mutated allele of a gene that affects connective tissue. Connective tissue is found in joints, the lens of the eye, the aorta—in other words, all over your body.

But having the Marfan syndrome doesn't mean that all those parts will be affected. I might still have the condition even though I have only some of the symptoms. The doctors started to figure this out after my mom died and my cousin Amber developed some problems with her eyes. Now they suspect that my grandfather, my uncle, and my cousin might have the Marfan syndrome. My grandfather and uncle are healthy so far, like me. The only way they can be sure if we have the Marfan syndrome is to test our genes. Dr. Foster says it's important to know if I have it so we know how closely to monitor my heart.

Because the Marfan syndrome can be caused by many different mutations in this gene, scientists can't test for just one Marfan mutation. They have to check to see if I inherited the mutated allele that runs in my family. They will compare my DNA to DNA from my grandfather, uncle, cousin, and other family members who do not have the Marfan syndrome. This test will let me know for sure if I have the Marfan gene that my mother had.

My dad still doesn't want me to be tested. He's worried people will treat me differently, or that he won't be able to get health insurance for me. Dr. Chee, my genetic counselor, said that can be a problem. She thinks more

➤

laws will be passed to prevent genetic discrimination. But, right now, we aren't sure if the legal system will protect me.

If the test is positive, I will probably have to quit playing soccer. We were hoping I'd get a soccer scholarship to college—Coach thinks there's a good chance I could. But I've been doing some reading. If you have the Marfan syndrome they keep a close watch on your heart. They can even do surgery if you develop serious heart problems. The doctor says I can probably live to a healthy old age, even if I do have this condition, as long as I take the right precautions.

What do you think I should do, Megan?

Joe

ANALYSIS

1. Joe's family pedigree is shown below. How would a genetic counselor answer the following questions?

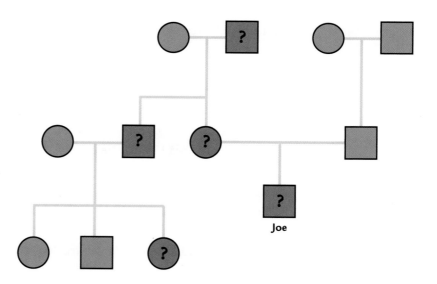

Joe

a. If Joe has the Marfan syndrome, is he likely to be homozygous or heterozygous? Explain your answer.

b. If Joe has the Marfan syndrome and has a child someday, what is the probability that his child will have the Marfan syndrome? (Assume that the child's mother does not have the Marfan syndrome.) Make a Punnett square and explain your answer.

2. The Marfan syndrome is a dominant trait. Write your own definition of *dominant trait* as it is used in genetics. Use evidence to explain whether the dominant trait is always the most common trait in a human population.

3. Pretend that you are Joe's friend. Write a letter to Joe telling him whether you think he should be tested.

Hint: To write a complete answer, first state your opinion. Provide two or more pieces of evidence that support your opinion. Then consider all sides of the issue and identify the trade-offs of your decision.

EXTENSION

For links to more information on genetic conditions, go to the SALI page of the SEPUP website. Use the website as a starting point to research issues related to genetic conditions and genetic testing.

INVESTIGATION

The study of genetics provides scientists with answers to many questions and provides people with some practical solutions to problems. You learned about some of the practical outcomes of genetics when you studied Joe's dilemma. Genetic testing can help people take action about genetic conditions, but there are concerns involved as well.

Another practical outcome of genetics research is the ability to identify people. For example, wars and political actions sometimes lead to young children being taken from their families. When many years have passed, how can families ever find their children again?

CHALLENGE

Can you use blood types to help identify lost children?

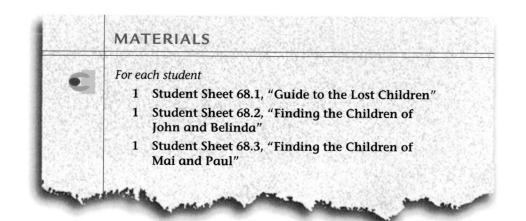

MATERIALS

For each student

1 Student Sheet 68.1, "Guide to the Lost Children"

1 Student Sheet 68.2, "Finding the Children of John and Belinda"

1 Student Sheet 68.3, "Finding the Children of Mai and Paul"

SEARCHING SCIENTIFICALLY

Belinda and John and their good friends, Mai and Paul, almost never talk about their lost children. Five years ago, they were separated from their young children when enemy troops invaded their village of Namelia. Belinda and John escaped with their oldest son, but their two younger children, Serena and Noah, were captured. Mai and Paul lost both of their two young children. Many other young children and some of the elder people in the village were also captured.

When peace returned, the villagers searched for their lost children. After months of searching and waiting, there was no sign of them. But recently, the local authorities helped one family find one of their children in the village of Samarra, 100 miles away. He had been 6 years old when he was taken, so he remembered his parents and his name. Belinda and John decide to travel to Samarra with Mai and Paul.

When they reach Samarra, the authorities tell them they have found some children who were adopted during the war five years ago. These children were adopted shortly after the attack on Namelia, and evidence suggests they were originally from that village. The citizens of Samarra had fought the enemy troops and were able to save some of the children. The families who adopted the children have taken good care of them and they do not want to give them up. Nevertheless, the authorities hire some genetics experts, including you, to help identify the lost children.

One way to begin the process of identifying children of specific parents is to look at their blood groups. The ABO blood group and Rh factor will be used to determine which children cannot be matches. Rh factor is another kind of blood type. Rh positive is dominant over Rh negative.

PROCEDURE

1. You have been chosen to join the genetics team going to Samarra. Before you leave you must review what you have learned about the genetics of the ABO blood group. Meet in your group of four to do this.

2. To help you with your assignment, look at your copy of Student Sheet 68.1, "Guide to the Lost Children." Fill in the possible alleles for each person in Tables 1 and 2.

3. Examine the information in Tables 1 and 2 about Belinda and John, Mai and Paul, and the eight children who are the right ages to be their children. Use this information to complete the steps below to decide which children might belong to each of the two couples.

    a. First determine which of the girls might be Belinda and John's lost daughter Serena. Use the table provided on Student Sheet 68.2, "Finding the Children of John and Belinda," to explain your reasoning for each of the four girls.

    b. Determine which of the boys might be Belinda and John's lost son Noah. Use the table provided to explain your reasoning for each of the four boys.

    c. Determine which of the boys might be Mai and Paul's lost son Ben. Use the table provided on Student Sheet 68.3, "Finding the Children of Mai and Paul," to explain your reasoning.

    d. Determine which of the girls might be Mai and Paul's lost daughter Jade. Use the table provided to explain your reasoning.

ANALYSIS

1. How certain are you that some of the eight children belong to Belinda and John or to Mai and Paul?

2. What additional evidence would help you identify the lost children?

MODELING

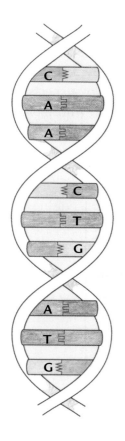

DNA is double-stranded. One strand provides the information in the gene. Both strands are needed when the gene is copied.

Blood type results show that some of the lost children *might* be Belinda and John's or Mai and Paul's, but how can the investigators know for sure? DNA typing can be used to check for exact DNA matches. This is sometimes called **DNA fingerprinting** because it gives a unique result that helps identify people, but it is actually very different from regular fingerprinting. Since DNA fingerprints of relatives are much more alike than those of unrelated people, they can be used to find out if people are related.

Each person, except for identical twins, has unique genetic information. This information is encoded in long molecules of DNA in the chromosomes. DNA can be extracted from cells, cut into pieces, sorted, and stained. The pattern of these DNA fragments looks almost like a complicated bar code. DNA fingerprinting reveals your own unique pattern, which is almost as unique as your DNA itself.

The genetic code is made up of four "letters" (A, T, C, and G), each of which stands for one of four related chemicals that are strung together in the DNA. The order of these letters provides information. Since the sequences of the genes do not vary much among people, fingerprinting genes would not easily give information that could be used to tell people apart. (Two different alleles of the same gene might vary in only 1 of hundreds of letters.) However, long regions of DNA between the genes vary a lot more among people and can be used to tell people apart. That is why we "fingerprint" these regions between the genes.

CHALLENGE

What makes your DNA fingerprint unique?

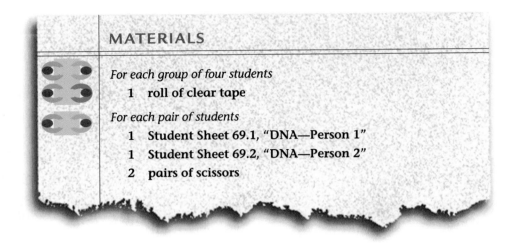

MATERIALS

For each group of four students
   1   roll of clear tape

For each pair of students
   1   Student Sheet 69.1, "DNA—Person 1"
   1   Student Sheet 69.2, "DNA—Person 2"
   2   pairs of scissors

PROCEDURE

Part One: How are DNA fingerprints used?

1. Compare the DNA fingerprints in the illustration. Blood found on a broken window pane at the scene of a burglary provided the first fingerprint. The other two fingerprints are from two suspects. In your science notebook, record your conclusion about whose blood was on the window.

Different lengths of DNA are in different positions on the "fingerprint." That's because the fragments of a person's DNA are sorted by length. Note that some of the bands are darker than others. Pieces that are the same length are piled in one spot. Therefore, the dark bands represent the more common lengths of DNA in each person.

Part Two: Why is every person's DNA fingerprint so different?

2. Take one of the Student Sheets showing DNA sequences and give the other to your partner. Keep the Person 1 DNA separate from the Person 2 DNA.

3. Assemble the DNA strand: Cut out the strips of DNA. Tape them together in order, forming a single long strand. (Real DNA extracted from cells has several billions of these letters—and they're already strung together!)

4. Hold the DNA from Person 1 above the DNA from Person 2. Compare the DNA sequences, looking for similarities and differences. Record your observations in your science notebook.

5. Cut the DNA strand by making cuts only AFTER the sequence of letters AAG. For example, the following sequence would be cut like this:

A A T G C T A G T T C G G T G A T A A G ✂ C G G T G T T A G G C T T A G

6. Sort your DNA pieces, with the shortest piece at the right end of the desk and the longest piece at the left end of the desk. See the example below.

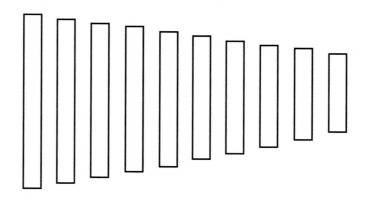

7. Read "How DNA Fingerprinting Is Performed in the Lab" on next page.

How DNA Fingerprinting Is Performed in the Lab

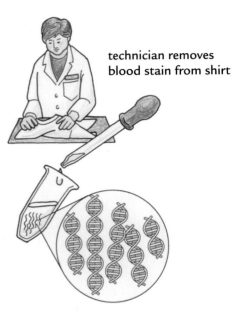

technician removes blood stain from shirt

1. Technicians put the cell sample into a tiny test tube. They break open the cells and use chemicals to extract, or take out, the DNA.

2. The DNA is cut into pieces with special chemicals called enzymes.

3. The pieces of DNA are separated based on their length:

   • The pieces are put into a hole in a flat, rectangular agar gel.

   • The technician then runs an electrical current across the gel, which makes the pieces of DNA move toward the positive end.

   • The smaller pieces of DNA move faster than the bigger pieces.

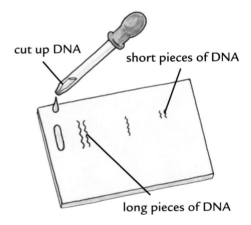

cut up DNA

short pieces of DNA

long pieces of DNA

4. Special techniques are used to make the bands of differently sized DNA visible.

ANALYSIS

1. In your science notebook, create a table like the one below. In the table, match the steps you did in the simulation to the steps scientists use to make DNA fingerprints.

What scientists do	What we did in the simulation
Extract DNA from cells	
Cut the DNA with enzymes	
Use an agar gel and electric current to separate DNA pieces	
Make the DNA visible	

2. Look at this DNA fingerprint.

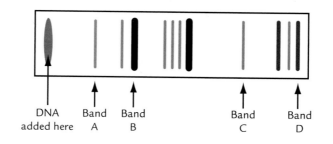

a. Which single band represents the smallest pieces of DNA? Explain how you can tell.

b. Which single band represents the most common length of DNA for this fingerprint? Explain how you can tell.

3. Why are DNA fingerprints unique to each person? In your explanation, refer to the way that DNA is cut up and sorted, and refer to the DNA of Person 1 and Person 2 from the activity.

EXTENSION

Go to the SALI page of the SEPUP website for links to websites about the Human Genome Project. Here you can explore some of the latest research on human genes.

DNA testing is more time-consuming and expensive than blood typing; that's why the investigators of the lost children used blood typing first. But DNA fingerprints are so reliable and show so much about a person's genetic information that they provide very strong evidence in cases involving identification.

DNA fingerprinting is used to verify biological relatives, to identify bodies, and to figure out who was present at the scene of a crime. Special techniques have recently been developed to increase the DNA found in very tiny specimens: DNA can now be collected from microscopic specks of blood, fragments of old teeth, or even the root of a single hair!

Many scientists use DNA fingerprinting and similar techniques in their work. Anthropologists use it to study ancient remains such as Egyptian mummies. Biologists use DNA fingerprinting to compare the DNA of various living organisms in order to investigate evolutionary relationships. Paleontologists have even used the technique on dinosaur blood found in insects preserved in amber. (But dinosaur cloning is still far from being possible!)

CHALLENGE

How can DNA fingerprinting help find the lost children?

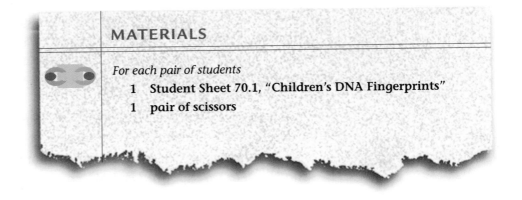

MATERIALS

For each pair of students

1 Student Sheet 70.1, "Children's DNA Fingerprints"
1 pair of scissors

PROCEDURE

1. Figure 1 shows the DNA fingerprints of a child, his or her biological parents, and an unrelated child. Examine and compare the DNA fingerprints. Record your observations in your science notebook.

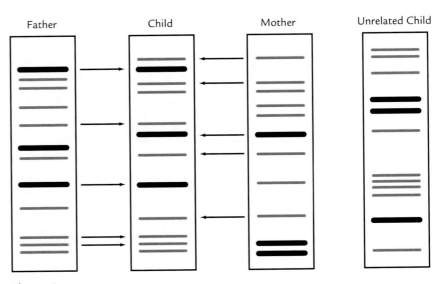

Figure 1

2. To try to identify their missing children, Mai, Paul, Belinda, and John each give blood and have their DNA fingerprints made. They obtain a court order to test the blood of the children in Samarra.

Student Sheet 70.1, "Children's DNA Fingerprints," has the DNA fingerprints for the eight children you considered in Activity 68, "Searching for the Lost Children." Cut out the DNA fingerprints and compare them one by one to the parents' DNA fingerprints, shown in Figure 2 below. Be sure to include the label with the fingerprint (for example, "Girl 1").

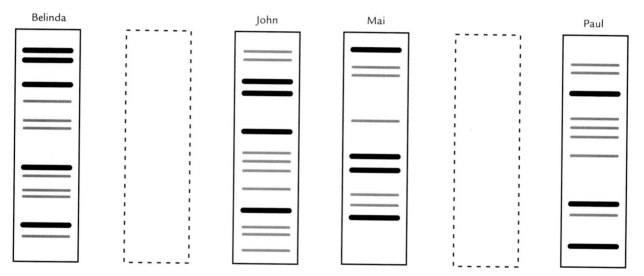

Figure 2

3. Record your observations and conclusions in your science notebook.

4. Luckily, Belinda remembers that she kept her daughter Serena's baby tooth that was knocked out when she was two years old. The root of the tooth is used to extract some of the daughter's DNA. Compare the daughter's DNA fingerprint to the DNA fingerprints of the girls in Samarra. Record your observations.

Serena's DNA fingerprint

ANALYSIS

1. Use DNA fingerprint evidence and the blood type evidence from Activity 68, "Searching for the Lost Children," to explain each of the following:

   a. Which child or children are not likely to be those of Belinda and John?

   b. Which child or children are likely to be those of Belinda and John?

   c. Which child or children are not likely to be those of Mai and Paul?

   d. Which child or children are likely to be those of Mai and Paul?

2. Write a convincing statement about which of the eight children (if any) are the children of Belinda and John, and which of the children (if any) are the children of Mai and Paul. In your statement, provide as much evidence as you can to convince a judge that the biological children of these parents have been found. Be sure to include evidence from previous activities.

TALKING IT OVER

DNA fingerprinting can be used to identify children who are lost or kidnapped at a young age. Dr. Mary-Claire King helped develop and apply genetic techniques to identify over 50 lost children in Argentina. The methods she developed continue to help parents and grandparents find children in other countries torn by war and political conflict.

These people in Argentina are demanding to know what happened to the children who disappeared.

CHALLENGE

What are the ethical issues involved in using genetic information?

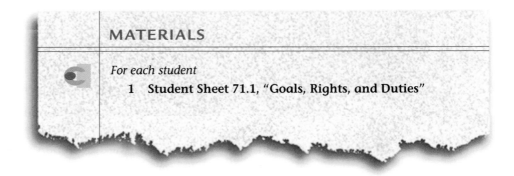

MATERIALS

For each student
1 Student Sheet 71.1, "Goals, Rights, and Duties"

A TRUE STORY

From 1976 to 1983, a military dictatorship controlled the country of Argentina. During that time, at least 15,000 citizens who were opposed to the dictatorship "disappeared." Most of the adults were killed. Witnesses reported that most of their children who were too young to talk or remember what happened were taken away to live with other families. At the end of the dictatorship, about 200 of these children were thought to still be alive.

The grandmothers of the lost babies formed a group and collected stories about the missing children. As the children entered kindergarten, school secretaries would secretly phone the grandmothers to report suspicious-looking birth certificates from newly registered students. Other evidence was also available: some of the mothers arrested during the dictatorship had managed to write the names of their children and themselves on prison walls before they died. In 1983, a new government replaced the dictatorship and a presidential commission was formed to study the problem. Thousands of people were interviewed. All these strands of evidence were collected and woven together. By 1984 the grandmothers thought they knew where many of their grandchildren were, but more evidence was needed before disrupting the lives of the children and their adoptive families. A genetics expert named Dr. Mary-Claire King was consulted.

Dr. King had perfected an advanced DNA fingerprinting technique to help make identification of biological relationships more reliable. She also helped set up a collection of blood samples and DNA fingerprints.

Using several types of evidence (including genetic evidence), some of the children were identified. Even with sophisticated genetic techniques, not all families were reunited: of the 200 missing children, only 50 were found by their grandmothers. Yet Dr. King's work continues to help people in Argentina and in other areas of the world.

Cases like this raise an important ethical question. Scientists can find ways to identify lost children, but how should society use this information? The decision to return the children to their biological parents affects adoptive parents, biological parents, and children. Sometimes the answers to these questions are difficult.

MAKING A DECISION

Consider the story of the lost children of Namelia. The court is convinced that Belinda and John have found both of their children. Mai and Paul have found their son. But the families who adopted the children were guilty of no wrong. They adopted the orphaned children, have taken good care of them, and want to keep them just as strongly as their biological parents want them back. A group of ethics experts has been asked to recommend what should be done.

You will use methods of ethical analysis to analyze the case of the lost children of Namelia. An analysis of goals, rights, and duties will help you explore the ethics of this case.

PROCEDURE

Think about the goals, rights, and duties of each group of people involved in the decision of whether to return the children to their biological parents. Record your ideas on Student Sheet 71.1, "Goals, Rights, and Duties."

ANALYZING GOALS, RIGHTS, AND DUTIES

Goals: *What is the action intended to accomplish?*

In this case, what goals would be accomplished by returning the children to their biological parents or grandparents? Goals may differ depending on whom you ask.

For example, the goals of a member of a sports team participating in a game might include having fun, making friends, and/or winning. The goals of the coaches may be the same, or may include other goals, such as teaching good sportsmanship. The goal of the umpire may be to be certain the game is played fairly and according to the rules. The goal of a parent may be for the team member to get exercise.

Rights: *What are the rights of the people involved?*

For example, in the United States, the right to an education, the right to freedom, and the right to a trial if one is accused of a crime are just a few of the rights to which everyone is entitled by law.

Duties: *What are the duties of the people involved?*

For example, parents have a duty to provide food and a home for their young children. A judge in a courtroom has a duty to ensure that the rules of the court and the law are followed.

ANALYSIS

1. What are the pros and cons of reuniting the children with their biological parents or grandparents seven years later?

2. Imagine you are a judge trying to make a fair and final decision about whether to reunite the children from Samarra with their biological families from Namelia. Write your ruling and your explanation. Be sure to discuss any difficult ethical trade-offs you have had to make.

   Hint: To write a complete answer, first state your opinion. Provide two or more pieces of evidence to support your opinion. Then consider all sides of the issue and identify the trade-offs of your decision.

3. **Reflection:** How does the goals, rights, and duties method help you think about ethical issues?

EXTENSION

Research an issue in genetics that interests you. To get started with links to more information on recent research and issues in genetics, go to the SALI page of the SEPUP website.

Ecology

E

Unit E

Ecology

Michael loved to bike through the park. The air smelled fresher there than on the street, and he always saw so many interesting things.

Once, he had come across a bird's nest with several young chicks still in it. As he watched, one of the parents had brought food for the chicks to eat. He wondered if he would see anything like that today.

Suddenly, he saw a small frog near the edge of Turtle Pond. It looked very familiar. In fact, it looked just like the frog his sister kept as a pet. It was different from the frogs he usually saw at Turtle Pond.

"Could that be my sister's frog?" Michael wondered. If so, how did it get there? Did it escape, or could his sister have let it go? Could a pet frog survive in Turtle Pond? How would it affect the other animals that also lived in the pond?

• • •

What are the relationships between an organism and its environment? What effect do humans have on these relationships?

In this unit, you will explore ecology: the study of the relationships between organisms, including humans, and the environment.

TALKING IT OVER

Have you ever thought that it would be cool to have parrots flying around in your backyard? Or wished that there were hippos in your local lake? What happens when you introduce an organism into a new environment?

FISHING ON LAKE VICTORIA

James Abila is a Kenyan boy of 17. His family has a small fishing boat on Lake Victoria. He sat outside his hut to talk to us. Inside, his mother was preparing lunch, while his sister and younger brother were laying out a few fish to dry in the afternoon sun.

James started his story. "My father made our boat. He was always one of the best fishermen in the village. He still catches all kinds of fish, though he says it's not as easy as it used to be. Most of the fish in the lake used to be very small, just 2–4 inches long. So it was easy to use our net to catch hundreds of small fish. But about the time I was born, the number of fish seemed to go down. Luckily, the government introduced new fish into the lake. Now, the most common fish in the lake is Nile perch. It's a much bigger fish and can be too heavy to catch with a net. That's why I work for one of the fishing companies. They have the large boats needed to catch Nile perch. And I can earn money to help feed my family."

Uganda
Kenya
Lake Victoria
Tanzania

CHALLENGE

What are the trade-offs of introducing a species into a new environment?

PROCEDURE

Work with your group to read and discuss the story of Nile perch in Lake Victoria.

NILE PERCH

Lake Victoria is the second largest lake in the world and it contains some extremely large fish. One type of fish found there, known as Nile perch *(Lates niloticus),* can grow to 240 kilograms (530 pounds), though its average size is 3–6 kilograms (7–13 pounds). But Nile perch weren't always found in Lake Victoria. Until the 1980s, the most common fish in Lake Victoria were cichlids (SICK-lids), small freshwater fish about 2–4 inches long. (If you've ever seen aquarium fish such as oscars, Jack Dempseys, or freshwater angelfish, you've seen a cichlid.)

This man is holding a large Nile perch.

Lake Victoria cichlids interest **ecologists**—scientists who study relationships between organisms and environments—because there are so many species of these fish. Although they all belong to the same family (see Figure 1), at one time there were over 300 different species of cichlids in Lake Victoria. Almost 99% of these species could not be found anywhere else in the world!

Figure 1: Classification of Cichlids	
Kingdom	Animalia
Phylum	Chordata
Class	Osteichthyes (bony fish)
Family	Cichlidae

There used to be many other kinds of fish in the lake, including catfish, carp, and lungfish. The 30 million people who lived around Lake Victoria relied on the lake for food. Because most of the fish were small, they could be caught by using simple fishing nets and a canoe. The fish were then dried in the sun and sold locally.

By the late 1950s, however, it appeared the lake was being overfished. So many fish were caught that the populations remaining did not have enough members left to reproduce and grow. If the lake continued to be overfished, there might not be enough fish left for people to eat. As a result, the British government (which ruled this part of Africa

Cichlids are one of the many small fish commonly found in Lake Victoria.

at that time) decided to introduce new fish species, such as Nile perch, into the lake. They wanted to increase the amount of fish that was available to eat; they hoped to provide more high-protein fish for local people and to be able to sell extra fish to other countries. Ecologists were opposed to this idea. They were worried that the introduction of Nile perch, which had no natural enemies within the lake, would negatively affect the lake's ecosystem. Before a final decision could be made, Nile perch were secretly added into the lake. Eventually, more Nile perch were deliberately added by the government in the early 1960s.

During the 1960s and 1970s, before there were a lot of Nile perch in the lake, about 100,000 metric tons of fish (including cichlids) were caught each year. By 1989, the total catch of fish from Lake Victoria had increased to 500,000 metric tons. Today, each of the three countries surrounding the lake (Uganda, Kenya, and Tanzania) sells extra fish to other countries. In Figure 2, you can see how the amount of fish caught by Kenyan fisheries has changed over a 15-year period.

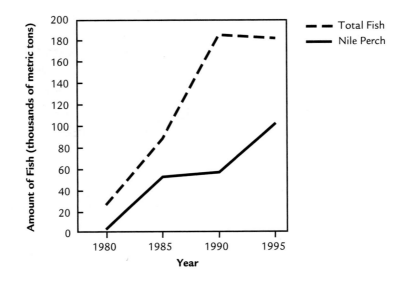

Figure 2: Amount of Fish Caught in Lake Victoria by Kenya

Besides increasing the amount of fish, there have been other consequences of introducing Nile perch into the lake. Because Nile perch are large and eat other fish, they are believed to have caused the extinction of as many as 200 species of cichlids. The populations of other types of fish, including catfish and lungfish, have also declined. Many ecologists are upset that their predictions have come true.

Some of the cichlids that have become extinct ate algae. With their extinction, the amount of algae in the lake has increased 5-fold. Algae use up a lot of oxygen, making it difficult for other tiny plants and animals to survive in the lake. Today, many of the deeper parts of the lake are considered "dead" because they don't contain much living matter.

However, many of the original goals have been met. In 1979, there were 16,000 fishermen along the Kenyan shores of the lake. In 1993, there were 82,300. Many people are now employed by companies that process and sell Nile perch overseas. Over time, these fish have brought more money into the African countries surrounding the lake. Local people, who now eat Nile perch as part of their diet, consider Nile perch a "savior."

Some ecologists wonder how long the current situation can last. Nile perch are predators. As populations of other fish decline, the Nile perch's food sources are declining. The stomachs of some large Nile perch have been found to contain smaller, juvenile Nile perch. What will happen to the population of Nile perch if their food supply dwindles even further? Will the Nile perch population be overfished like the fish populations before it? Only time will tell.

ANALYSIS

1. Based on the reading, how did the amount of fish caught in Lake Victoria change from the 1960s to 1989?

2. Based on Figure 2, describe how the amount of Nile perch caught by Kenya changed from 1980 to 1995.

3. Look again at Figure 2. How do you think the number of metric tons of fish caught relates to the size of the total fish population from year to year? Explain your reasoning.

4. How did the introduction of Nile perch affect the food supply of the people who lived near Lake Victoria?

5. What effect did the introduction of Nile perch have on the organisms that lived in the lake?

6. Should Nile perch have been introduced into Lake Victoria? Support your answer with evidence and discuss the trade-offs of your decision.

   Hint: To write a complete answer, first state your opinion. Provide two or more pieces of evidence that support your opinion. Then consider all sides of the issue and identify the trade-offs of your decision.

7. What do you predict will happen to Lake Victoria over the next 20–30 years? Why?

PROJECT

ntroduced, non-native, exotic, and *non-indigenous* are all words used to describe species that humans have introduced outside of the species' normal range. The Nile perch is an **introduced species** that was placed deliberately into Lake Victoria. In other cases, the introduction of a new species into a new environment is accidental. Consider the case of the zebra mussel, which is named for the black and white stripes found on its shell. It was accidentally introduced into the United States in the 1980s and it is now estimated to cause up to $5 billion dollars of damage each year!

CHALLENGE

What effect can an introduced species have on an environment? What, if anything, should be done to control introduced species?

MATERIALS

For the class
    books, magazines, CD-ROMs, Internet access, etc.

For each student
  1   Student Sheet 73.1, "Introduced Species Research"

PROCEDURE

1. Read about the introduced species described on the following pages. As directed by your teacher, decide which one species your group will research.

2. Over the next few days or weeks, find information on this species from books, magazines, CD-ROMs, the Internet, and/or interviews. You can also go to the SALI page of the SEPUP website to link to sites with more information on species mentioned in this activity.

3. Use this information to complete Student Sheet 73.1, "Introduced Species Research." You should provide the following:

- common and scientific name of your species

- its native and current range; its relationship to and effect on people

- its effect on new ecosystem(s)

- the reasons for its success

- issues related to its future growth or spread.

Later in this unit, you will use your research to create a class presentation.

EXTENSION

Visit a local greenhouse or botanical garden. Look at the labels of ornamental plants used in landscaping. Where did these plants originally come from? Is the introduction of these species considered to be good or bad?

Kudzu Brings Down Power Lines!

Kudzu (KUD-zoo), sometimes referred to as "the vine that ate the South," has finally pushed local patience to the limit. Properly called *Pueraria lobata*, it was first introduced in the 1920s to the southern United States as food for farm animals and to reduce soil erosion. Today, this fast-growing vine from Japan has overgrown entire forests and choked local ecosystems. Last week, the weight of kudzu vines pulled down power lines, causing a two-day power outage. Mayor Lam has called for control measures. All community members are invited to a town council meeting to consider what should be done to control this destructive vine.

Response to Tiger Mosquitoes Raises Questions

The public outcry over the worsening problem with the tiger mosquito *(Aedes albopictus)* continues. In response, the city has begun nighttime spraying of insecticide. Jesse Butler, principal of the Little Town Preschool, said, "How can the city be allowed to spray poison on the backyards where children play?" City spokesperson Kate O'Neil told reporters that the insecticide is harmless to people. "Tiger mosquitoes are very aggressive. They are much worse than the native mosquitoes. Apart from the nuisance, tiger mosquitoes can spread diseases such as yellow fever. We have to take action!" O'Neil invites interested residents to attend the Camford Mosquito Abatement Board presentation on the tiger mosquito problem and possible solutions.

Nutria Hunting on State Marshes?

Ecologists from City University are considering teaming with local hunters in a surprise move to reduce the population of nutria (NEW-tree-uh) in state marshes. Nutria *(Myocastor coypus)* are large, beaver-like rodents whose burrows and voracious grazing are causing serious damage to marshes.

Ecologist Charlie Desmond told reporters that nutria are native to South America.

They were brought to North America for their fur. When they escaped into the wild, their population exploded. "If we don't act soon, we could lose our marshlands in just a few years," he cautioned. Duck hunters, bird watchers, sport fishers, and hikers are pressuring the state legislature to come up with a solution. Nutria hunting is one option being seriously explored.

Aquarium Plant Turns Out to Be Worst Weed

You may have seen this aquatic plant sold in small bunches at aquarium stores. It's a popular plant because goldfish like swimming between its stems. But when aquariums are dumped out into lakes, ponds, or rivers, hydrilla (hie-DRILL-uh) can quickly grow into a dense mat that chokes out other vegetation. This change of the environment is dramatic for native animals and plants. *Hydrilla verticillata*, as it is known scientifically, can clog up city water intake valves and get tangled in boat propellers. "We used to have the best swimming hole down by the bridge," said Rita Aziz, a 7th grader at Junior Middle School. "Now it's filled with this gross weed. The last time I swam there, I got tangled in it. It was scary. I would really like to find a way to do something about it."

Cut Down Trees to Protect Them?
Agency Advises on Longhorn Beetle Threat

When Keesha Murray, age 3, was injured by a falling branch in Tot Play Park, local neighborhoods woke up to the threat of the Asian longhorn beetle. Her father, Toby Murray, said that Keesha had played under the big maple tree many times. Under the attack of the Asian longhorn beetle, the tree had recently died, which led to the loss of the tree limb. "Keesha was scratched up and scared. We were lucky it wasn't worse," he said.

Shade trees all over the city have been dying due to the recent invasion of this wood-boring beetle from Japan, known scientifically as *Anoplophora glabripennis.* The beetle larvae are very hard to kill. One suggestion is to cut down all trees within city parks to prevent the beetle from spreading.

A Landscape Beauty Is Taking Over

What is the link between landscaping your yard and the recent reports that local marsh species are declining? Purple

loosestrife *(Lythrum salicaria),* whose magenta flowers are admired by gardeners, is the weed to blame. It was introduced from Europe as a medicinal herb in the early 1800s and is still sold today as a landscaping plant. According to the Fish and Wildlife Service ecologist Johanna Brown, "It totally takes over an area, crowding out native species. It's really devastating for fragile marsh ecosystems." Brian Van Horn, a teacher at Middleton Junior High, is also concerned. "It's a tough plant to get rid of and killing it can damage the marshes even more." A meeting at Middleton Junior High will be held to discuss this issue.

Farmers Rally to Scare Off Starlings

The recent outbreak of hog cholera may be related to starling *(Sturnus vulgaris)* droppings getting into pig food. Carol Polsky, a pig farmer in Poseyville, encouraged local farmers to work together to help get rid of the birds. "In addition to spreading disease, those birds eat crops, seeds, and animal feed. A flock of starlings will eat just about anything and they poop everywhere. That spreads disease to other animals, not just pigs," Polsky told reporters.

Many control options are available, according to Dr. Tony Caro of the Agricultural Sciences Board. Dr. Caro commented, "In 1891, 60 starlings were released in New York and now they are the most common bird in America!" But a representative of the local nature society told reporters that the latest annual survey showed that starling populations had dropped since the previous year. Dr. Caro will be speaking at the next meeting of the County Farm Association, where control measures for starlings will be discussed.

Brown Snake Problem Bites Guam

Guam, a tiny, tropical island, is a U.S. territory with a problem. People have been bitten. Bird, bat, and lizard populations have declined. The culprit? The brown tree snake *(Boiga irregularis)* from New Guinea.

After baby Oscar Gonzalez was bitten by a brown tree snake, local people were spurred to action. "Most of us know about them. Those snakes climb the power poles and short out electricity on the island several times a week," Nicki DeLeon, a long-time resident of Guam, told reporters. "Back in the 1960s and even the 1970s, the jungle was full of birds singing. We used to see bats and little lizards running around. They're not so easy to find now."

Scientists are working to find ways to control the snake before the last of the unique island species disappear forever. Dr. Sheila Dutt, a researcher with EcoSave International, said, "As well as helping with snake control on Guam, we are desperate to prevent this snake from hitching a ride in air cargo. I don't even want to think of the effect this snake could have in other parts of the United States."

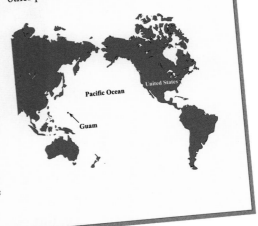

LABORATORY

How do scientists know how introduced species affect ecosystems? Natural environments are constantly changing. How do you figure out what changes are due to the introduced organism and what changes are due to other factors? **Ecology** is the study of relationships between living organisms and the physical environment. Ecologists begin by studying organisms in the natural environment. They often supplement this information with laboratory investigations.

CHALLENGE ➔

What can you discover about an organism in a laboratory investigation?

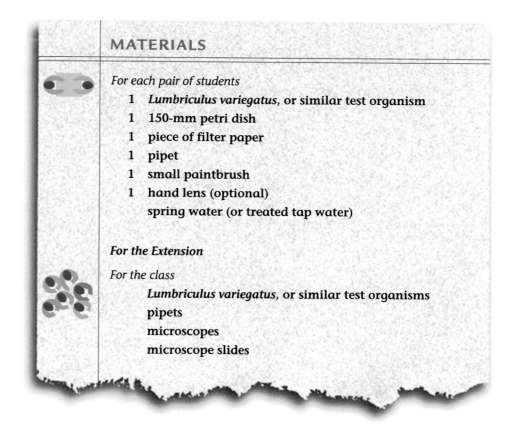

MATERIALS

For each pair of students

1   *Lumbriculus variegatus,* or similar test organism
1   150-mm petri dish
1   piece of filter paper
1   pipet
1   small paintbrush
1   hand lens (optional)
    spring water (or treated tap water)

For the Extension

For the class

Lumbriculus variegatus, or similar test organisms
pipets
microscopes
microscope slides

PROCEDURE

1. Pour 1–2 cm of water into the bottom of the petri dish.

2. Use the pipet to add a single blackworm from the culture to your petri dish. (Do not pick a blackworm that is dark and has a lighter section at one or both ends; this worm has recently been broken and is regenerating itself.)

3. Carefully observe the blackworm. Then use the brush to gently investigate this organism.

4. Record your observations. For example, how does the blackworm move? Does it respond differently to different actions, such as touching? Can you identify which end of the blackworm is the head? What else do you observe about the blackworm?

5. Place the filter paper in the lid of the petri dish. Use the pipet and a few drops of water to completely moisten the filter paper.

6. Use the pipet to move the blackworm onto the filter paper.

7. Observe the blackworm's movement on this surface. How does its movement here compare with its movement in water? Record your observations.

8. Return your blackworm to the class culture before cleaning up.

EXTENSION

Place a blackworm on a microscope slide. Add one drop of water. (If there is too much water on the slide, use a pipet to suction off the excess water. Use your finger, not your mouth, to suction the water.) Observe the worm under low and medium power. What internal structures can you see?

ANALYSIS

1. Review your notes on how the blackworm responded to touch. How could these reactions help it to survive in the wild?

2. Based on what you now know about blackworms, in what type of environment do you think blackworms live? Explain your reasoning.

3. As an ecologist, you are asked to write an entry in an encyclopedia on the blackworm, *Lumbriculus variegatus.* Use your laboratory notes to write a paragraph describing the blackworm.

4. **a.** A student reading your encyclopedia entry thinks that you should include more information about blackworms. What questions do you think he or she might have after reading your entry?

   b. How might you get the information necessary to answer his or her questions?

There are many types of introduced species—just think about the differences between starlings and purple loosestrife! Most of the well-known cases belong to the plant or animal kingdom. While you may recognize kudzu, loosestrife, and hydrilla as plants, you may not have realized that all of the other introduced species discussed so far, including zebra mussels and tiger mosquitoes, are part of the animal kingdom. In fact, there are over one million known animal species in the world today, with many more being discovered every year. With such a large diversity of species, how do you know if the animal you are studying is similar to one another scientist is studying?

The five-kingdom classification scheme, shown in Figure 1, is one way of classifying species based on observations of their structures and other characteristics. This system helps scientists group species together to make sense of the diversity of life. It allows scientists to compare an organism, such as a zebra mussel, to other organisms with similar characteristics. In this activity, you will focus on organisms found in the animal kingdom.

The five-kingdom classification scheme is one of several ways scientists classify organisms. Other approaches to classification will be introduced in activities in the Evolution unit.

Figure 1: The Five-Kingdom Classification Scheme

Animals	Plants	Fungi	Protists	Bacteria

CHALLENGE

What are some similarities and differences among animals?

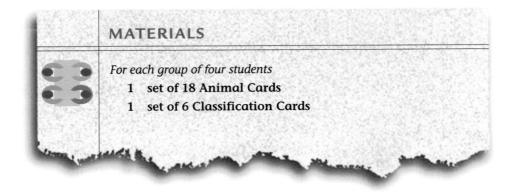

MATERIALS

For each group of four students
1   set of 18 Animal Cards
1   set of 6 Classification Cards

PROCEDURE

Part One: Exploring the Animal Kingdom

1. Spread your Animal Cards out on a table.

2. Look at each of the Animal Cards, noting similarities and differences among the animals.

3. Read the information on each card. This information represents what you might discover if you observed the animals more closely and were able to dissect a specimen.

4. With your student group, classify the animals into 4–8 groups. Work together to agree on a classification system:

   • If you disagree with others in your group about how to classify an animal, explain to the rest of the group why you disagree.

   • Listen to and consider other people's explanations and ideas.

5. In your science notebook, describe groups that you created: How many groups did you create? What do the animals in each group have in common? Be sure to record which animals you placed in which group.

6. Share your categories with another group of students. Explain why you classified the animals the way you did. Discuss how your group's categories were similar to or different from those of the other student group.

Part Two: A Biologist's Perspective

7. Spread out the Classification Cards on the table. (You may remember these cards from a previous *Science and Life Issues* activity.) Use the cards to review the characteristics of the five kingdoms and the non-living viruses.

8. Biologists use information similar to the descriptions provided on the Animal Cards to divide kingdoms into large categories called **phyla** (FIE-luh). The singular of *phyla* is *phylum*. Each phylum contains similar species. There are about 35 animal phyla. Your teacher will share with you how biologists group the animals on your cards into six of these phlya. Humans are grouped in the phylum Chordata, as shown below.

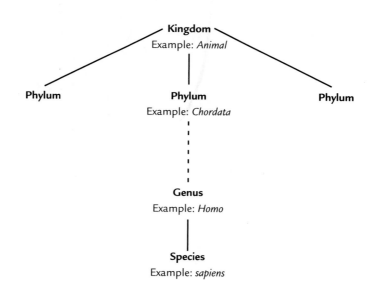

Humans are members of one of many phyla of animals.

9. Adjust your animal groups so they look like the phyla used by biologists today. Then complete Analysis Questions 1–3.

ANALYSIS

1. How did your categories change when you followed the biologists' system of phyla? Did your number of categories increase, decrease, or stay the same?

2. Look carefully at how biologists group these animals into phyla. What types of characteristics are used to group animals into phyla?

3. Animals without backbones are called invertebrates. How many invertebrate phyla do the animals on your Animal Cards represent? List these phyla.

4. **Reflection:** What characteristics were most important to you when you grouped the Animal Cards? How are these characteristics different from the ones that biologists use to classify? What do you now think is the best way to group animals? Explain.

INVESTIGATION

One of the 35 animal phyla—phylum Chordata—includes all species with backbones. Most of the chordates have a jointed backbone and are classified in the sub-phylum Vertebrata, or vertebrates. Although only about 50,000 vertebrate species have been identified (compared to about 1 million invertebrate species), the most familiar animals are vertebrates, such as humans, elephants, eagles, and frogs. How are vertebrates classified into smaller groups?

CHALLENGE

What kinds of evidence can you use to classify vertebrates?

PROCEDURE

1. Carefully read the Five-Kingdom Classification Chart on pages E30–E31 to compare characteristics of the five classes of vertebrates. "Cold-blooded" animals are animals that adjust their body temperature by moving to warmer or cooler locations. Their temperatures sometimes vary with the environment's temperature, but they aren't always cold. "Warm-blooded" animals regulate their body temperature to a fairly constant level by generating heat within their bodies, but they aren't always warm. Because of this, scientists now use different terms to describe these animals.

2. Pretend you work at a zoo. Some people have discovered some strange vertebrates and ask you for help in identifying them. They have sent you letters containing pictures and descriptions of these creatures. You can find the letters on pages E25–E29.

   For each mystery vertebrate:

   a. Read the letter and look at the picture.

   b. Discuss with your group members which vertebrate class might include this species.

   c. In your science notebook, record which class you believe it belongs to and your reasons. You do not need to agree with your group members.

ANALYSIS

1. What characteristics do you think best distinguish each vertebrate class?

2. Why do some vertebrates appear to fit into two or more different classes?

My husband and I were having lunch outside at our hotel in Mexico when I saw a small creature flash across the wall. I later saw a similar animal sunning itself outdoors. I'm enclosing a picture. The next day, I managed to catch one. It was sunning itself on a rock and its skin felt hot and dry, not moist. I could feel a line of bones along its back. As I held it, it seemed to get a little stressed; I noticed that it started to breathe faster. So I set it down and it ran off. We really liked these creatures and would like one as a pet. What kinds of animals are these?

J. Stirbridge

One of my kindergarten students brought in a picture of this animal. Hariette told the class that she saw one of these animals when she lived in New Zealand. She said that it looked hairy and that it was very rare. Harriette and her dad saw the animal poke around for worms with its sharp beak. Her dad is out of the country and Harriette wants to do a project on this animal. What is it? Thank You.

Mr. Kalmus and Class K-1

May 23, 1860

My collecting party was recently in the new territory of Australia, where we were astounded to find a most amazing variety of strange and unknown animals. The animal I have sketched below appears to be truly new to the world of science. We have also made observations of these creatures in their natural habitat. They live in ponds and streams and are covered by dark fur. The animal has a bill like a duck, which it uses to find snails and food in the mud of the stream. We then observed something most extraordinary. The female lays an egg which she keeps in her pouch until it hatches. The tiny baby licks milk from the skin of the mother's belly.

What is your opinion of these mysterious new creatures?

Sincerely,

Murray Jones

MY girlfriend and I accidentally ran over this thing on our last road trip! Melia ran over to pick up the animal as soon as I stopped. The animal looked scaly, but had some hairs poking out between the "scales." Although it was a cold night, Melia said its body was still warm. Melia wants to put up signs warning people to look out for these animals so that no one else accidentally runs one over, but we don't know what they are. She's an artist, so she drew a picture of it for you. Can you help us identify this animal?

Tim

Nina and I are in 5th grade. We love to go snorkeling near the reefs by my house in Guam. We saw some very strange-looking animals underwater. I tried to draw one for you. They have a head like a horse but they have a fin on their back. One day, we saw one of them moving around and then some babies came out near its stomach! The babies swam straight to the surface but then came back down. We watched and watched but never saw them go back up to the surface. How can they breathe? What are they?

Thanks, Thomaso

I was scuba diving in Thailand when I saw this long, striped creature, maybe as thick as my thumb, working its way along the bottom and sticking its head into holes. Its head was smaller than an eel's and I know that eels stay in their holes during the day. (This was a day dive.) Also, the animal was smooth and round, with no fins. I also noticed that it regularly went to the surface for air. Any ideas about what it is?

Phil

I am writing to ask you about some flying animals that nearly flew right into us when Pearl and I went caving last summer. We were near the entrance to a cave when I heard this twittering sound and saw some shadows fly past me. Pearl panicked and ran. She wouldn't go back to the cave. Later that night, I went exploring myself. When I shone my flashlight on the ceiling, I saw hundreds of really tiny animals hanging there. They seemed to be grouped together to keep warm because the cave was so cold. I think they were babies, because they looked much smaller than the creatures I had seen before. I saw one of the larger creatures fly into the cave and go to one of the babies. The baby seemed to be getting milk from the adult. I was wondering if you could help me figure out what these things are.

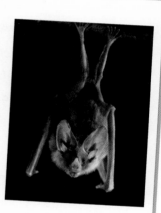

Sincerely yours,

Thelma

From: Ruby Riter

Subject: strange animal

I'm a travel writer with the Leisure Time Gazette. I was on assignment in Malaysia and saw these strange animals on the mud near mangrove swamps. I want to write about them for next week's travel section, but I need more information. I saw some of these animals swimming underwater, but I didn't see any of them come up for air. However, they seemed to do okay on land too. When I checked them out through a telephoto lens, I noticed that they had some kind of fin going down their back as well as scales on their bodies. Can you get back to me ASAP? My deadline is in three days. Thanks a lot.

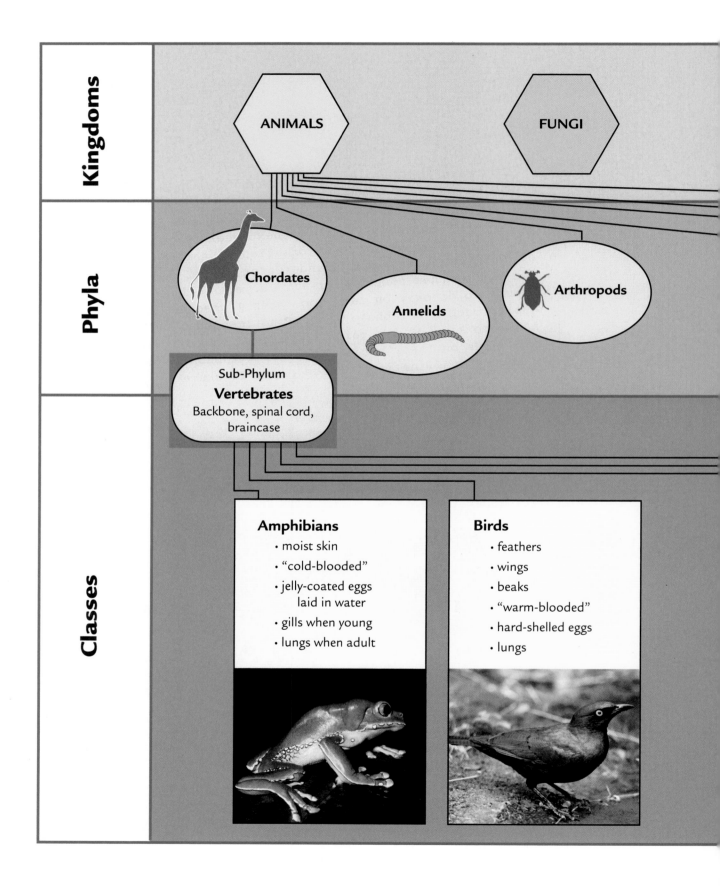

Kingdoms

ANIMALS

FUNGI

Phyla

Chordates

Annelids

Arthropods

Sub-Phylum
Vertebrates
Backbone, spinal cord,
braincase

Classes

Amphibians
- moist skin
- "cold-blooded"
- jelly-coated eggs
  laid in water
- gills when young
- lungs when adult

Birds
- feathers
- wings
- beaks
- "warm-blooded"
- hard-shelled eggs
- lungs

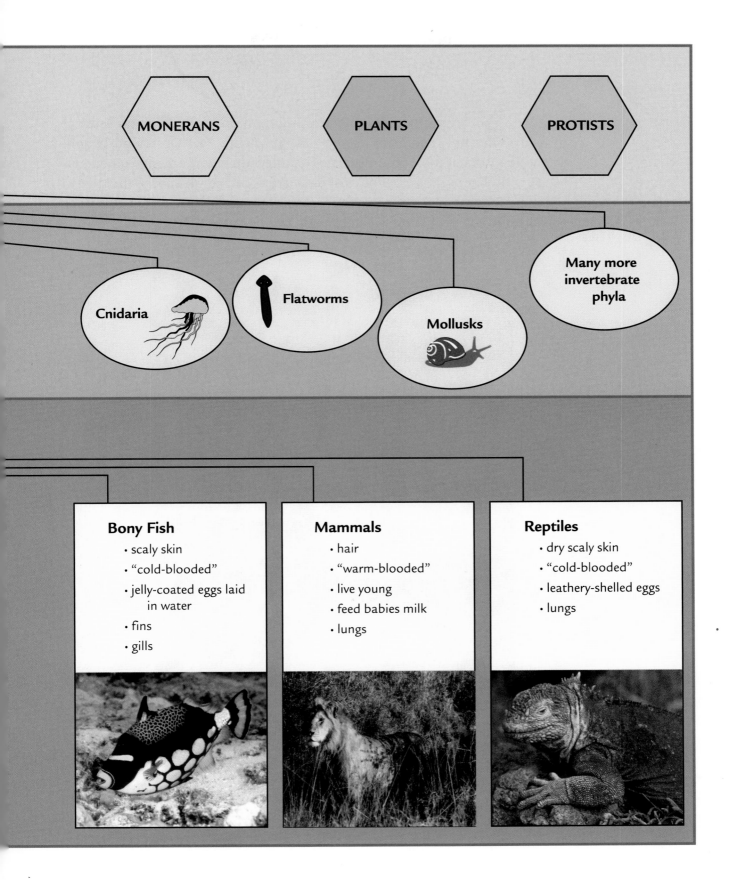

MONERANS   PLANTS   PROTISTS

Cnidaria

Flatworms

Mollusks

Many more invertebrate phyla

Bony Fish
- scaly skin
- "cold-blooded"
- jelly-coated eggs laid in water
- fins
- gills

Mammals
- hair
- "warm-blooded"
- live young
- feed babies milk
- lungs

Reptiles
- dry scaly skin
- "cold-blooded"
- leathery-shelled eggs
- lungs

INVESTIGATION

You can gather ecological information by studying an individual organism, as you did in Activity 74, "Observing Organisms." But most organisms do not affect an environment as individuals, but as groups. Groups of individuals of a single species that live in the same place are known as populations. The photos on this page and the next show different populations of sea lions.

This population of sea lions lives on piers in a harbor.

One introduced species that is causing a lot of problems in the United States is the zebra mussel. Its success in freshwater environments has caused the loss of native wildlife as well as damage to equipment. How fast is this population spreading? Some investigators predict that populations of zebra mussels will be found across the entire United States within 20 years. Studying what has happened to populations of zebra mussels in lakes around the world can help scientists figure out what changes are occurring in the U.S. and what to expect for the future.

CHALLENGE

How do scientists study the size of a population and predict future population changes?

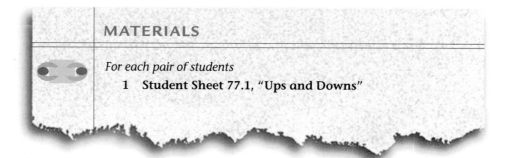

MATERIALS

For each pair of students
1 **Student Sheet 77.1, "Ups and Downs"**

This population of sea lions lives on a beach.

PROCEDURE

Part One: Initial Observations

1. In your group of four, review the two tables below. Imagine that two different groups of ecologists collected data on the size of the zebra mussel population in Lake Miko for two different time periods.

Table 1: Zebra Mussel Population in Lake Miko, Period 1 (1959 to 1968)				
Year	1959	1960	1962	1968
Number of Zebra Mussels (per square meter)	2,211	95	93	97

Table 2: Zebra Mussel Population in Lake Miko, Period 2 (1971 to 1976)				
Year	1971	1972	1974	1976
Number of Zebra Mussels (per square meter)	393	802	1,086	2,179

2. Divide your group in half. Assign one of the two data tables to each pair within your group.

3. With your partner, create a line graph of the data in your table using Student Sheet 77.1, "Ups and Downs." Remember, independent variables, such as time, are always graphed on the x-axis. Since you will compare graphs within your group, make sure that the x-axes of both graphs use the same scale.

4. After completing your graph, respond to the two questions on Student Sheet 77.1 as directed.

Part Two: A More Complete Analysis

5. Show your graph to the other students in your group. Point out the overall population trend—is the population increasing, decreasing, or staying the same?

6. Compare the two graphs. Discuss what conclusions you can make about the population trend in Lake Miko during Period 1 vs. Period 2.

7. Place the two graphs together, with the graph for Table 1 first and the graph for Table 2 second. If necessary, fold the edges of your sheets to fit the graphs together.

8. As a group, discuss what happens to the population trends when the two graphs are connected. Discuss how what you see with the two graphs together is different from what you see with each of the individual graphs. Be sure to:

   • Describe what happens to the population size of zebra mussels in Lake Miko from 1959–1976.

   • Discuss whether you can make any definite conclusions about whether the population is increasing, decreasing, or staying the same.

ANALYSIS

1. **a.** Sketch a line on your graph predicting what you think will happen to the size of this population of zebra mussels during the ten years after 1976.

   b. Explain your prediction. Why do you think the graph will look that way?

   c. What additional information would make you more confident of your prediction? Explain.

2. **a.** What factors do you think affect the size of a population?

   b. Explain how each factor might affect population size: Would it cause the population to increase, decrease, or stay the same? Why?

3. As you know from your own graph, data were not collected every year. Explain whether you would expect a well-designed experiment to collect data every year. What might prevent the collection of such data?

4. Shown below are graphs of zebra mussel populations in three lakes near Lake Mikolajskie. Describe the population trend in each graph. How does each population change over time?

Zebra Mussel Populations in Three Lakes

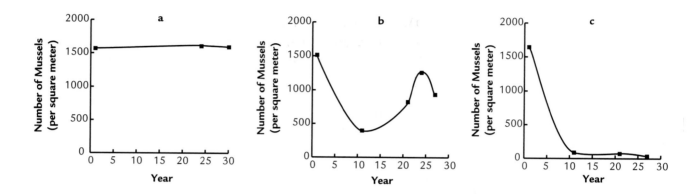

5. The data presented in this activity are similar to actual data collected in Lake Mikolajskie, Poland, between 1959 and 1987. Zebra mussels have been found in lakes in that area for over 150 years. Shown below are the data collected from 1977–87. How does this additional information compare to your answer to Question 1?

Table 3: Zebra Mussel Population in Lake Miko, Period 3 (1977 to 1987)					
Year	1977	1979	1982	1983	1987
Number of Zebra Mussels (per square meter)	77	104	81	55	85

6. Zebra mussels were introduced in the United States in the late 1980s. They first appeared in Lake Erie, one of the Great Lakes. Today, the population of zebra mussels has reached as high as 70,000 mussels per square meter in some parts of Lake Erie.

   a. How does this compare to the populations of zebra mussels found in the lakes in Poland?

   b. Before 1988, the population of zebra mussels in Lake Erie was zero. Draw a graph showing what you think the data might look like for the population of zebra mussels in Lake Erie from 1985 to the present.

7. Consider the zebra mussel population in Lake Mikolajskie from 1959 to 1987. Describe what you think happened to the zebra mussel population from 1987 to 1997. Explain your reasons for your prediction.

LABORATORY

How do introduced species affect other organisms within a habitat? What happens to the populations of native species when a new organism is introduced? Scientists often draw diagrams, called **food webs**, to model the feeding relationships within an ecosystem. By showing what each organism eats, food webs model the energy relationships among species.

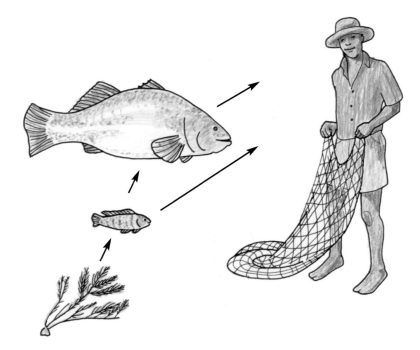

A simplified food web of Lake Victoria

How can you find out what an organism eats? One way is to examine its stomach contents. But in the case of owls, you can also examine an owl pellet. An owl pellet is a combination of bones and fur that an owl coughs up, just as a cat coughs up a hairball. Owl pellets are formed when owls swallow their prey whole and their digestive system cannot break down fur and bones. Within 12–24 hours after eating, an owl throws up a pellet. Piles of pellets are often found at the base of the tree on which an owl is perched. These pellets help ecologists learn what and how much owls eat.

CHALLENGE

What can an owl pellet tell you about an owl's diet? How can you use this information to develop part of a food web?

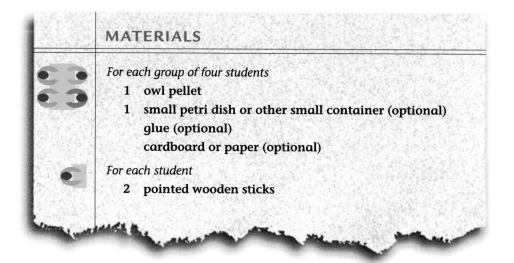

MATERIALS

For each group of four students
1   owl pellet
1   small petri dish or other small container (optional)
    glue (optional)
    cardboard or paper (optional)

For each student
2   pointed wooden sticks

PROCEDURE

1. Use the wooden sticks to carefully pull the owl pellet into four equal-sized pieces. Provide each member of your group with one of the four pieces.

2. Use your pair of sticks to gently separate all of the bones from the fur of your piece of owl pellet.

3. Work with your group to divide all of the bones into groups based on their shapes. Use Table 1, "Guide to Owl Pellet Bones," to help you.

4. Count and record the number of bones in each of your categories.

5. Try to arrange the bones to make a skeleton of one (or more) animal. Sketch your final arrangement(s).

➢

Table 1: Guide to Owl Pellet Bones

Skulls	
Jaws	
Shoulder blades	
Front legs	
Hips	
Hind legs	
Assorted ribs	
Assorted vertebrae	

ANALYSIS

1. What did you learn about the diet of owls from investigating an owl pellet? Include information about the type and number of organisms in an owl's diet. (Remember that an owl ejects a pellet within 12 to 24 hours after eating.)

2. **a.** The organisms that you uncovered in your owl pellet are likely to be voles, small rodents similar to mice. Owls also eat other small mammals, such as shrews, and insects. Use this information on owl diet to develop a food web.

   b. Voles eat mostly plant material such as grass, seeds, roots, and bark. Shrews eat insects. Add these relationships to your food web.

   c. The great horned owl sometimes eats other owls. It also eats small mammals like voles. Add the great horned owl to your food web.

3. Copy the graph shown below, which is similar to graphs you made in Activity 77, "Ups and Downs." It predicts the change in the population of owls as they first move into a new habitat.

Owl Population Over Time

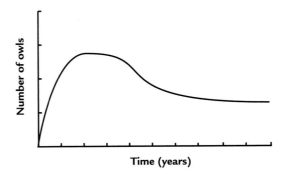

a. Draw a line showing what you think will happen over the same time period to the population of one of the species that owls eat.

b. Draw a line, using a different color or symbol, showing what you think might happen over the same time period to the population of one of the species that eats owls. Be sure to include a key identifying what species each line represents.

EXTENSION

Research the food web of the introduced species you are studying. What effects, if any, has your species had on native species? What effects do you predict it will have in the future?

One important part of every organism's habitat is a source of food. The introduction of new species into an **ecosystem** often changes the availability of food.

CHALLENGE

How are the energy relationships among organisms in an ecosystem affected by the introduction of a new species?

READING

Is it possible that a scenario like the one in Lake Victoria could happen in the United States? Scientists are waiting to see. But in the United States, the main concern isn't a large predator like the Nile perch, but a seemingly unimportant mussel less than two inches long. The tiny zebra mussel *(Dreissena polymorpha)* (see Figure 1) doesn't seem large enough to cause serious problems. But its ability to reproduce and spread quickly is making it into a big issue.

Figure 1: Zebra Mussels Feeding

Zebra mussels reproduce by releasing eggs and sperm into the water. The fertilized eggs grow into tiny larvae. Because of their small size, they are very hard to see at this stage.

STOPPING TO THINK 1

Brainstorm ways in which zebra mussels might accidentally be spread from one lake to another.

Zebra mussels feed on some of the smallest members of the aquatic food chain: microscopic animals and plants known as **plankton** (PLANK-tun) (see Figure 2). (When discussing them in more detail, biologists usually use the words *zooplankton* [zoe-uh-PLANK-tun] for microscopic animals and *phytoplankton* [fie-toe-PLANK-tun] for microscopic plants.) Plankton are found throughout the water, from the very deepest part of a lake to the surface. They are the food for a variety of other organisms, including many kinds of fish. In addition, zooplankton eat phytoplankton. Thus, phytoplankton are at the base of many aquatic food chains.

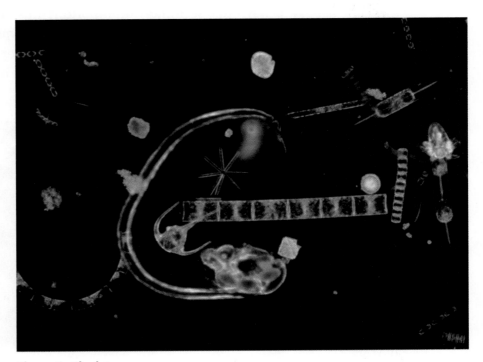

Figure 2: Plankton

Phytoplankton include microscopic plants and algae. These tiny organisms are especially important in aquatic ecosystems because they produce food for all the other living things in that ecosystem. You may know that plants and algae require sunlight in order to grow. They use sunlight as energy to convert carbon dioxide and water into food—a process known as **photosynthesis** (foe-toe-SIN-thuh-sis). (You will learn more about photosynthesis in Activity 81, "A Producer's Source of Energy.") The food that the plant produces is stored within the plant as starch or sugar. The plant can then use its food for activities within its own cells—until the plant is eaten by another organism! Since most plants and algae do not eat other organisms for food but are able to produce their own food, they are called **producers**. Producers such as phytoplankton form the base of the food chain because they have the ability to use the sun's energy to make their own food.

All other organisms rely on this ability of producers to convert the energy from the sun into food energy. Organisms that get their energy by eating food are known as **consumers**. Some consumers eat plants for energy, while other consumers eat the animals that eat plants. Some consumers, such as zebra mussels and humans, eat both plants and animals.

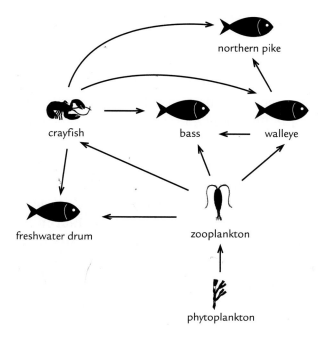

Figure 3: A Simplified Lake Food Web

STOPPING TO THINK 2

Why are producers, such as plants, an essential part of any ecosystem?

In Figure 3, you can see a simplified lake food web with both producers and consumers. The transfer of energy that takes place when one organism eats another is shown by arrows. Each arrow shows where the energy from the food is going within the ecosystem. The arrows show who is eaten by whom, not who eats whom. Many other species eat phytoplankton; food webs become more complicated when additional relationships are added.

STOPPING TO THINK 3

a. Copy Figure 3 into your science notebook. Identify each organism as either a producer or a consumer.

b. Think about the kinds of food that people eat. Use this knowledge to add humans into this lake food web.

c. In the lake food web, humans are consumers. Are humans always consumers? Explain.

After zebra mussels appeared in the Great Lakes ecosystem, they changed the food web. Zebra mussels filter water and catch the microscopic plankton that live in the water. They rely on phytoplankton and zooplankton for food. Because zebra mussels are often more common than other sources of food, crayfish and freshwater drum are starting to eat zebra mussels as part of their diet.

STOPPING TO THINK 4

Using Figure 3 as a guide, create a lake food web that includes zebra mussels. Be sure to show how zebra mussels get their energy *and* how other organisms get energy from them.

At first, these changes don't seem too important. After all, couldn't the lake ecosystem support one more consumer? Adult zebra mussels filter about one liter of water per day. This means that a two-inch mussel can filter enough water to fill half of a large soft drink bottle every day. In some parts of the Great Lakes, the concentration of zebra mussels has reached as high as 70,000 mussels in a square meter. This means that just a small area of mussels would be able to filter 70,000 liters of lake water each day! As a result of zebra mussels, the clearness of the water has changed: it is now 600% clearer than it was before the introduction of the zebra mussels. Clear water sounds like a good thing, but biologically speaking, extremely clear water can mean that there is not much alive in the water. In fact, the zebra mussel population has been so effective at filtering plankton that the populations of some types of phytoplankton have decreased by 80%.

Remember, phytoplankton are the base of this aquatic food chain. By removing large amounts of phytoplankton from the water, zebra mussels remove the food for microscopic zooplankton. Many types of fish depend on zooplankton for food. In some cases, these fish are the food for other fish and for humans and other mammals. Some ecologists predict that zebra mussels will change the entire food web of the lake ecosystem. However, there is no evidence yet that zebra mussels have affected fish populations in the lake.

There is evidence, though, that the types of plants in the lake are changing. Because of the increased clearness, sunlight is now able to penetrate deeper into the lake. Plants such as algae are now growing along the lake bottom. This provides habitat and food for other organisms, such as sunfish, that are currently not common in the lake. Some scientists predict that the fish populations will change: populations of some fish, like walleye, will decrease, while populations of other fish, like sunfish, will increase.

What will happen to the lake ecosystem? At this point, no one is sure. The one thing that everyone is sure of is that zebra mussels will spread. The dots in Figure 4 mark areas where the zebra mussel is now found. In the period from 2000–2005, the zebra mussel spread to Kansas, Nebraska, and Virginia.

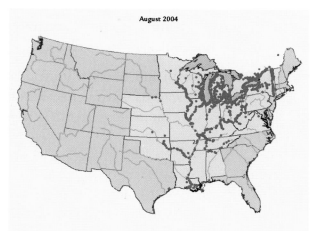

August 2004

Figure 4: Spread of Zebra Mussels Across the United States

STOPPING TO THINK 5

Look at Figure 4: the lines across the U.S. represent large rivers. Where do you predict zebra mussels will be found in the next 10 years? The next 20 years? The next 50 years? Explain your predictions.

EXTENSION

Go the SALI page of the SEPUP website to link to the website of the United States Geological Survey. What is the current status of zebra mussel spread across the U.S.?

ANALYSIS

1. A volcano erupts 40 miles from the lake ecosystem whose food web you drew in Stopping to Think 4. Ash from the eruption blocks sunlight over your ecosystem for several months. Explain what happens to each population within the lake food web in the weeks that follow the eruption.

2. The ash clears and several more months go by. Think about what is now happening to your lake ecosystem. Identify what factors will affect how quickly it recovers.

3. **Reflection:** Think about what you have learned about introduced species as well as ecosystems. What effect(s) can an introduced species have on an ecosystem?

LABORATORY

Y ou have learned about the roles of producers and consumers in a food web. But what about worms, bacteria, and fungi? What role do they play within an ecosystem? Organisms that eat dead organisms and wastes from living organisms are known as **decomposers.** Worms, bacteria, and fungi are decomposers. You can think of decomposers as a special type of consumer: they consume dead organisms and waste material.

Fungi such as these decompose wood and other dead plant material.

Decomposers like worms and bacteria can seem unimportant. The decay they cause can look (and smell) horrible. But decomposers are essential to the ability of ecosystems to recycle important nutrients like carbon and nitrogen. Decomposers like bacteria and fungi break down dead matter into chemicals that can be absorbed by plants. Without decomposers, dead organisms would pile up and the nutrients they contain could not be re-used by plants. Eventually, the fertility of soil and aquatic ecosystems would be reduced to nothing. Imagine what the bottom of a lake would look like without any decomposers!

CHALLENGE

Where can you find some decomposers? What do these decomposers look like?

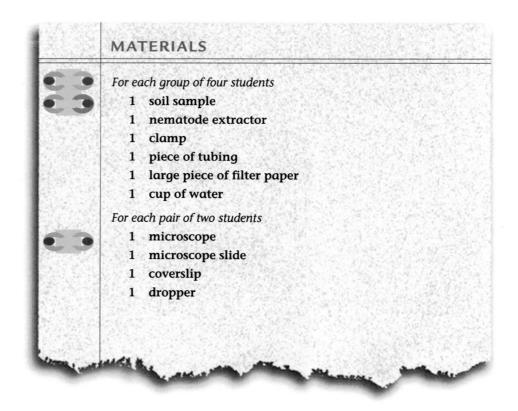

MATERIALS

For each group of four students

1   soil sample
1   nematode extractor
1   clamp
1   piece of tubing
1   large piece of filter paper
1   cup of water

For each pair of two students

1   microscope
1   microscope slide
1   coverslip
1   dropper

PROCEDURE

Part One: Investigating Soil

1. Gather ½ cup of soil from outdoors by scraping or shaking moist soil from around the roots of a clump of grass or other plant or from an area of decomposing leaf litter.

2. Place the tubing on the spout of the funnel. Then attach the clamp onto the middle of the tubing, as shown in Figure 1. Make sure that the tubing is pushed as far as it can go into the clamp; otherwise the water can drip out.

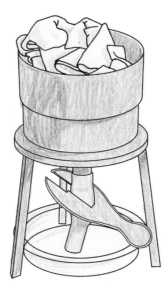

Figure 1: Nematode Extractor

3. Place the funnel in the stand and the perforated disc into the funnel.

4. Add water to the funnel to the level of the perforated disc.

5. Place the piece of filter paper in the funnel. Add a layer of your soil sample, no more than 1 cm deep, onto the filter paper.

6. Fold the filter paper over the soil. Add just enough water to cover the soil and filter paper. Set aside for one day.

Part Two: Searching for Nematodes

7. Carefully remove the clamp to release a small amount (less than 5 mL) of water into the cup. Share this sample in your group of four.

8. You might be able to see some small, white thread-like objects in the water. Try to suck up one of the thread-like objects into the dropper. Then squeeze a couple of drops from the dropper onto a microscope slide.

9. Carefully touch one edge of the coverslip, at an angle, to the mixture. Slowly allow the coverslip to drop into place.

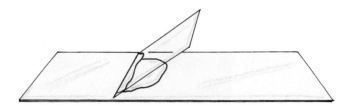

10. Begin by observing the slide on low power (usually the 4x objective). Be sure that the sample is in the center of the field of view (you may need to move the slide slightly) and completely in focus before going on to the next step.

    Hint: To check that you are focused on the sample, move the slide slightly while you look through the eyepiece—the sample that you are focused on should move as you move the slide.

11. Without moving the slide, switch to medium power (usually the 10x objective). Adjust the microscope settings as necessary.

    Hint: If material on the slide is too dark to see, increase the amount of light on the slide: do this by slightly opening the diaphragm under the stage.

$\succ$

12. While looking through the eyepiece, move the slide around slowly so that you see all parts of your sample. As you scan the slide, look for movement, especially of thin, colorless organisms like the ones shown in Figure 2. These organisms look like small earthworms, but are actually members of a different phylum. These tiny worms are called nematodes (NEM-uh-toads). (If you do not find any nematodes on your slide, make another slide from your sample.)

13. Try to count the number of nematodes on your slide. Compare the number of nematodes you and your partner find with the rest of your group.

14. When you have completed your observations, turn off the microscope light and set the microscope back to low power.

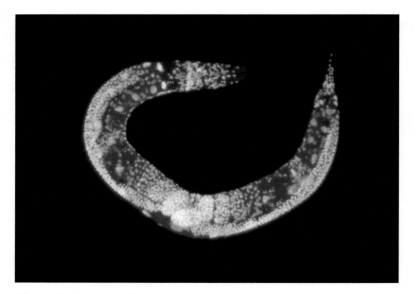

Figure 2: A nematode

ANALYSIS

1. Think about where some nematodes are found. What do you think they eat? Describe the role of nematodes in the ecosystem.

2. a. A simplified food web is shown on the next page. Which of the organisms in this ecosystem are producers? Which are consumers? Which are decomposers?

b. Use the food web to explain why decomposers could be considered a special type of consumer.

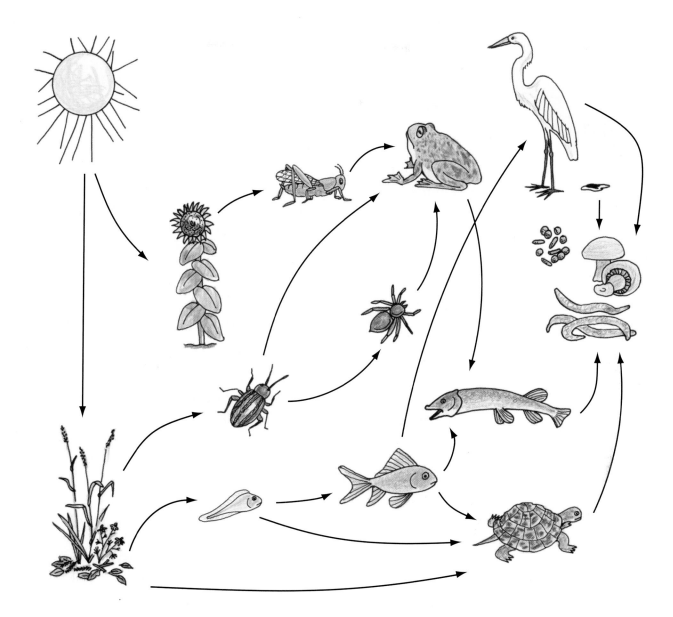

3. Like all organisms, birds like the egret need energy to live. Explain how the original source of energy for egrets, and all other consumers, is the sun.

4. Imagine that something kills most of the bacteria and other decomposers in a lake. What are some possible effects of killing these decomposers?

Producers such as phytoplankton and plants provide the energy for all other living creatures—for the consumers that eat plants, the consumers that eat animals that eat plants, and the decomposers that live off dead plants and animals. They do this by means of photosynthesis, a process by which plants use the energy from sunlight to convert carbon dioxide and water into food for themselves (and indirectly, for consumers). During this process, plants release oxygen gas into the atmosphere. Photosynthesis can be described by the following word equation:

$$\text{carbon dioxide} + \text{water} \xrightarrow{\textit{sunlight}} \text{food} + \text{oxygen}$$

Is light necessary for photosynthesis? How important is sunlight to an ecosystem? In this activity, you will use the indicator bromthymol blue (BTB) to collect evidence for the role of light in photosynthesis.

CHALLENGE

How do scientists study the role of light in photosynthesis?

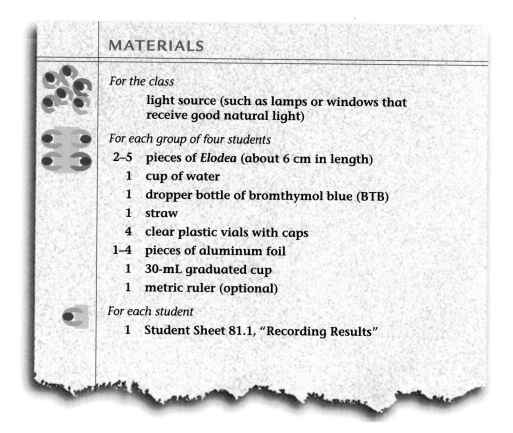

MATERIALS

For the class

light source (such as lamps or windows that receive good natural light)

For each group of four students

2–5	pieces of *Elodea* (about 6 cm in length)
1	cup of water
1	dropper bottle of bromthymol blue (BTB)
1	straw
4	clear plastic vials with caps
1–4	pieces of aluminum foil
1	30-mL graduated cup
1	metric ruler (optional)

For each student

1 Student Sheet 81.1, "Recording Results"

PROCEDURE

Part One: Collecting Evidence

1. If you have completed previous units of *Science and Life Issues,* review your notes from Activity 17, "Gas Exchange," and Activity 39, "Cells Alive!" Use your notes to complete Tables 1 and 2 on Student Sheet 81.1, "Recording Results." If you haven't completed these activities, your teacher will help you fill in the tables.

2. Fill a plastic cup half-full of water. (Your teacher may have already done this.) Add 15 drops of BTB.

3. Have one person in your group use a straw to blow into the BTB solution until it stops changing color. Record this as the initial BTB color in Table 3 of Student Sheet 81.1.

➤

4. Place a piece of *Elodea* into one of the vials. Carefully fill the rest of this vial with your BTB solution. Cap the vial tightly and place it in the light.

5. Fill a second vial with the same BTB solution only. Cap this vial tightly and place alongside the first vial.

6. With your group, discuss what you think might happen. Record your prediction in your science notebook.

7. After at least 45 minutes (or during your next class period), observe your vials. Use your observations to complete Table 3 of Student Sheet 81.1, as well as Analysis Questions 1 and 2.

Part Two: The Role of Light

8. Design an experiment to investigate the role of light in plant photosynthesis. **Hint:** Use the introduction to the activity and your results from Part One to help you.

When designing your experiment, think about the following questions:

- What is the purpose of your experiment?
- What variable are you testing?
- What variables will you keep the same?
- What is your hypothesis?
- How many trials will you conduct?
- Will you collect qualitative and/or quantitative data? How will these data help you to make a conclusion?
- How will you record these data?

9. Record your hypothesis and your planned experimental procedure in your science notebook.

10. Make a data table that has space for all the data you need to record. You will fill it in during your experiment.

11. Obtain your teacher's approval of your experiment.

12. Conduct your experiment and record your results.

EXTENSION 1

Observe a capped vial containing a plant in BTB solution at different times of the day. What color is the solution first thing in the morning? At lunchtime? Explain your observations. What process may be taking place in plants at night?

ANALYSIS

Part One: Collecting Evidence

1. What was the purpose of the vial containing only BTB solution?

2. In the introduction to this activity, you were told that plants need carbon dioxide during photosynthesis. What evidence do you have from Part One of your investigation to support this claim?

Part Two: The Role of Light

3. Describe your experimental results. Use the word equation on page E-54 to help explain your results.

4. Explain the role that light plays in photosynthesis. How do your results provide evidence for your explanation?

5. A second-grader comes up to you and says, "We just learned that the sun made all the stuff in my lunch. But my lunch was a tuna sandwich." Using language a second-grader would understand, explain how the sun was the original source of the energy in the tuna sandwich. Then try out your explanation on a child you know!

6. Think back to how the lake ecosystem described in Activity 79, "Eating for Energy," was affected by zebra mussels. Using your understanding of photosynthesis and ecosystems, explain why a decrease in phytoplankton allows more aquatic plants to grow on the lake bottom.

➤

EXTENSION 2

Your experiment looked at the *inputs* needed by a plant for photosynthesis. Design another experiment to collect evidence for the *outputs* of photosynthesis. Describe what materials you would need to perform this experiment, and what data you would collect.

As you have been learning, producers such as plants play a unique role within an ecosystem. By transferring the sun's energy into chemical energy stored in food, plants provide energy in a form that can be used by consumers and decomposers. What is different about plants that allows them to do this? Find out by investigating the cells of plants and then comparing them to animal cells.

A botanist (a person who studies plants) gathers plants for his research.

CHALLENGE

How are the cells of producers such as plants different from the cells of consumers such as animals? How do plant cell structures relate to their function as producers?

➤

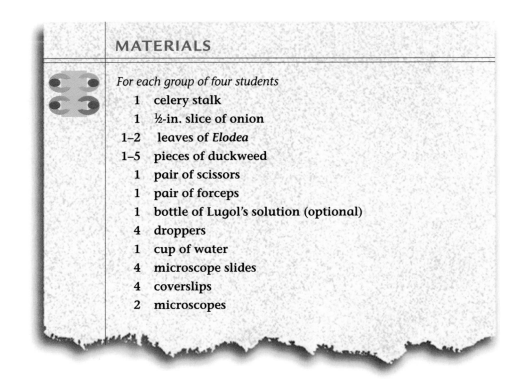

MATERIALS

For each group of four students

1 celery stalk
1 ½-in. slice of onion
1–2 leaves of *Elodea*
1–5 pieces of duckweed
1 pair of scissors
1 pair of forceps
1 bottle of Lugol's solution (optional)
4 droppers
1 cup of water
4 microscope slides
4 coverslips
2 microscopes

PROCEDURE

1. Have each person in your group complete one of the following four steps. You will share all four slides within your group.

a. Pull a string of celery off the stalk. At the edge of the string, you will see a thin film. This is the outer layer of the celery stalk and the part where you will see plant cells most clearly. Use scissors to cut a short length of this outer film. Place this piece of celery on a microscope slide. Add a drop of water and slowly drop the coverslip, at an angle, into place (as shown in Figure 1).

b. Get a small square of onion. Use your forceps to peel off a thin film of tissue from the inside layer of the onion square. Place this thin film on a microscope slide. Add a drop of water and slowly drop the coverslip, at an angle, into place (as shown in Figure 1).

c. Get a piece of *Elodea* and break off a leaf. Place a piece of this leaf on a microscope slide. Add a drop of water and slowly drop the coverslip, at an angle, into place (as shown in Figure 1).

d. Get a few leaves of duckweed. Place the leaves on a microscope slide. Add a drop of water and slowly drop the coverslip, at an angle, into place (as shown in Figure 1).

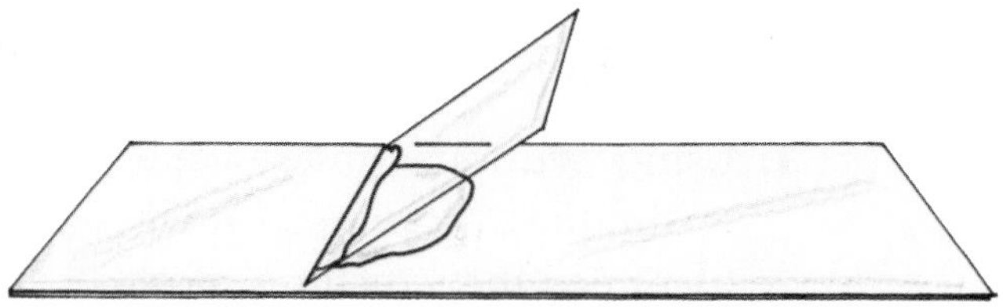

Figure 1: Placing the Coverslip

2. With your partner, observe the cells of each plant. Begin by observing the slide on low power (usually the 4x objective). Be sure that the plant material is in the center of the field of view (you may need to move the slide slightly) and completely in focus before going on to Step 3.

   Hint: When viewing celery, focus on the thinnest parts of the sample. When looking at duckweed, focus on the edges of the leaf.

3. Without moving the slide (which can be secured with stage clips), switch to medium power (usually the 10x objective). Adjust the microscope settings as necessary.

   Hint: If material on the slide is too dark to see, increase the amount of light on the slide: do this by slightly opening the diaphragm under the stage.

4. Turn the fine focus knob up and down just a little to reveal details of the plant cells at different levels of the slide.

5. Draw your observations of a cell from each plant. Be sure to record the type of plant and the level of magnification. Include details inside the cell and along the edge of the cell membrane on your drawing.

6. Look again at the duckweed, but this time look at a root. Draw your observations of the cells that you see in the duckweed root.

7. When you have completed your observations, turn off the microscope light and set the microscope back to low power.

EXTENSION

Place a drop of salt water at the edge of the coverslip while looking at *Elodea*. What happens? What does this tell you about the importance of fresh water to plants?

➤

ANALYSIS

1. Using various microscope techniques, scientists have identified the structures most commonly found in plant cells. Some of these structures are shown in the diagram of the plant cell shown in Figure 2. Not all plant cells contain every structure, though most plant cells do contain the majority of them. However, some of these structures are very difficult to observe if you only use a light microscope.

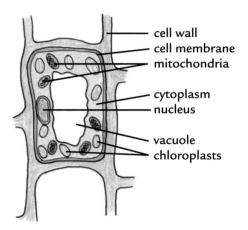

Figure 2: A Plant Cell

a. Which cell structures appear to be ones that you observed? List them.

b. Which cell structures were not visible to you? List them.

2. Compare the various plant cells you observed. Which cell structures did all of the plant cells appear to have in common?

3. Look at the simplified diagram of an animal cell shown in Figure 3. Animal cells, as well as plant cells, contain many structures; this diagram shows only some of these structures.

a. Which cell—plant or animal—is the cell of a consumer?

b. Compare the plant cell diagram with the animal cell diagram. Based on these diagrams, what structures would you expect to find in both plant and animal cells?

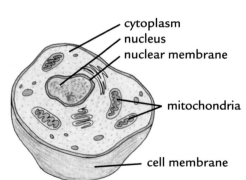

Figure 3: An Animal Cell

c. Based on your comparisons, which structure(s) within a plant cell do you think is most important in food production?

4. Copy a larger version of the Venn diagram shown here. Complete it by writing in the characteristics of animal cells, plant cells, and bacterial cells (which you may have first studied in Activity 44, "Who's Who?"). Record unique features of each type of cell in the individual spaces. Record common features among groups in the spaces that overlap.

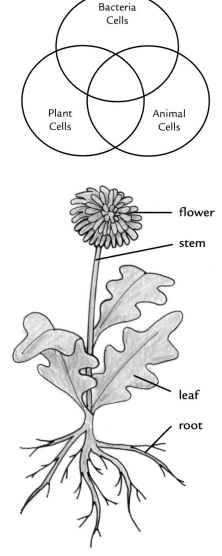

5. **a.** Many plants have leaves, stems, roots, and—during the blooming season—flowers. Which of these parts are likely to absorb sunlight and carry out photosynthesis?

b. Of the cells you observed—celery stem, onion, *Elodea* leaf, duckweed leaf, and duckweed root—which would you expect to carry out photosynthesis?

c. What cell structures are seen only in cells that absorb sunlight and carry out photosynthesis?

6. Three of the introduced species described in Activity 73, "Introduced Species," are plants: kudzu, purple loosestrife, and hydrilla. Each of these plants is growing successfully in different parts of the United States, partly because they are very well adapted to absorb sunlight and carry out photosynthesis.

a. What effect do you think the growth and spread of these introduced plants will have on native plants? Explain.

b. What effect do you think the growth and spread of these introduced plants will have on animals in the native ecosystems? Explain.

LABORATORY

Introduced species do not always survive in new environments. This is because all species have requirements for the place in which they can live. These requirements define the species' **habitat** (HAB-ih-tat). What makes up a habitat? Think about different aquatic ecosystems, such as a small pond or a coral reef. While both of these environments contain water, they have very different characteristics. Coral reefs are found in the ocean, which contains salt water, while most ponds are freshwater. An organism that lives in freshwater, like a zebra mussel, cannot survive in the coral reef environment. The photos below show several different habitats.

Producers, consumers, and decomposers are the living components of an ecosystem. Every ecosystem also has many non-living elements, such as rainfall, light, and temperature. The interaction of all these determines whether a habitat is suitable for a specific organism.

CHALLENGE

What are some of the important non-living characteristics of a habitat?

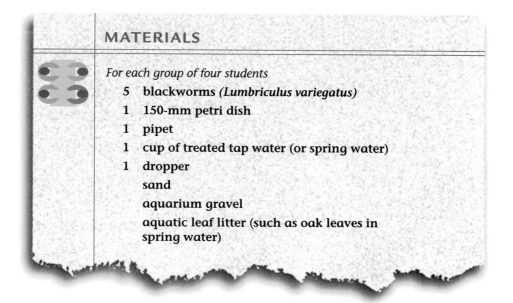

MATERIALS

For each group of four students

  5 blackworms *(Lumbriculus variegatus)*
  1 150-mm petri dish
  1 pipet
  1 cup of treated tap water (or spring water)
  1 dropper
     sand
     aquarium gravel
     aquatic leaf litter (such as oak leaves in spring water)

PROCEDURE

1. Fill the base of a petri dish with treated tap water (or spring water) and place 5 blackworms in it.

2. Observe how the blackworms respond over the next few minutes. Discuss with your group any behaviors that seem to be true of all or most of the blackworms.

3. As a class, discuss what type of data you could collect on the blackworms in order to determine which type(s) of material provides a good habitat for them.

4. Compare the different materials you can use to create a blackworm habitat. Record any similarities and differences in the physical characteristics of the different habitat materials.

➣

5. With your group, design an experiment to investigate which type(s) of material provides a good blackworm habitat.

When designing your experiment, think about the following questions:

- What is the purpose of your experiment?

- What variable are you testing?

- What variables will you keep the same?

- What is your hypothesis?

- How many trials will you conduct?

- Will you collect qualitative and/or quantitative data? How will these data help you to make a conclusion?

- How will you record these data?

6. Record your hypothesis and your planned experimental procedure in your science notebook.

7. Make a data table that has space for all the data you need to record. You will fill it in during your experiment.

8. Obtain your teacher's approval of your experiment.

9. Conduct your experiment and record your results.

ANALYSIS

1. Based on your experiment, which type(s) of material provides a good habitat for blackworms? Explain how your experimental results support your conclusions.

2. Describe the non-living characteristics of a habitat. **Hint:** What non-living factors could affect whether organisms will survive and reproduce?

3. What could you do with your blackworms to investigate if a warm or cold habitat is better for them? Write a procedure that anyone in your class could follow to investigate this question.

4. **Reflection:** Do you think that introduced species are always successful in new environments? Explain.

MODELING

Populations usually vary from season to season and year to year, often depending on non-living factors such as rainfall or temperature variations. Populations of a species can also be affected by living factors, such as other species that may provide food, compete for food, or provide shade or shelter.

When a new species is introduced into an area, it can compete with native species for food and other resources. Clams and zebra mussels are both mollusks that feed by filtering plankton from the water. What happens when zebra mussels are introduced into a habitat containing a clam population?

Zebra mussels growing on a native clam

CHALLENGE

How might the introduction of a competing species, such as zebra mussels, affect a population of native clams?

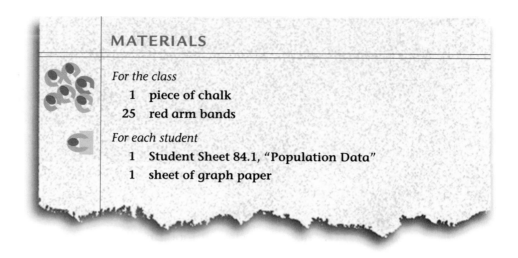

MATERIALS

For the class
  1   piece of chalk
  25 red arm bands

For each student
  1   Student Sheet 84.1, "Population Data"
  1   sheet of graph paper

PROCEDURE

Part One: Clam Population Size

1. As directed by your teacher, determine which students will initially represent clams and which students will initially represent plankton.

2. If you represent a clam, stand inside a chalk circle. There should be only one clam per circle. The space between the clams represents the amount of space a clam needs to survive. As long as you represent a clam, you must stay inside the circle.

3. If you represent plankton, stand behind the safety line on one side of the clam bed.

4. Your teacher will instruct the plankton to run through the clam bed, from one safety zone to the other (see Figure 1). A clam can use only one hand to tag its food. Each clam will try to "catch" (tag) plankton to survive; any plankton that is caught becomes a clam and has to find a home circle. Any clam that does not catch any plankton dies from lack of food; the student becomes plankton and must go to the safety zone.

5. Count and record the total population of clams.

6. Repeat Steps 4 and 5 at least ten times.

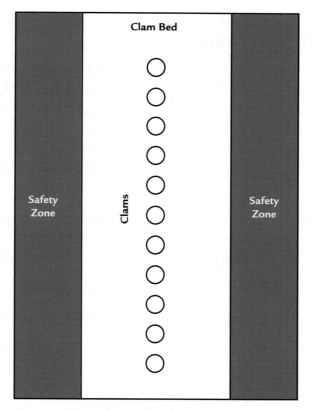

Figure 1: Map of Clam Catch Game

Part Two: Competition

7. Zebra mussels have invaded the clam bed! As directed by your teacher, determine which students will initially represent clams, which students will initially represent plankton, and which students will initially represent zebra mussels.

8. If you represent a zebra mussel, wear an arm band to identify yourself and then stand inside a chalk circle. Since zebra mussels grow very close together, a zebra mussel can grow in (i.e. share) the same circle as a clam. If no clams are present, two zebra mussels can occupy the same circle. As long as you represent a zebra mussel, you must stay inside a circle.

9. If you represent a clam, stand inside a chalk circle. There can still be only one clam per circle (although one zebra mussel can occupy the same circle). As long as you represent a clam, you must stay inside the circle.

10. If you represent plankton, stand behind the safety line on one side of the clam bed.

11. Your teacher will instruct the plankton to run through the clam bed, from one safety zone to the other (see Figure 1). A clam can use only one hand to tag its food, while a zebra mussel can use both hands. Each clam and zebra mussel will try to catch plankton to survive; any plankton that is caught becomes a clam or a zebra mussel (depending on who catches it). If you become a zebra mussel, collect an arm band to wear.

Any clam or zebra mussel that does not catch any plankton dies from lack of food and becomes plankton. When a zebra mussel dies, the arm band should be removed.

12. Record and count the total population of clams and zebra mussels.

13. Repeat Steps 11 and 12 at least ten times.

14. Record the class data on Student Sheet 88.1, "Population Data."

EXTENSION

Are Introduced Species Always Successful?

Introduce a mobile predator that eats only clams. Figure out how to modify the game to include this predator. Predict what you think will happen to the predator population and the clam population over time. Then test your ideas by playing the game for at least ten rounds.

ANALYSIS

Part One: Clam Population

1. a. Graph the population of clams over time from Part One of the Procedure. Decide which type of graph (bar or line) would best represent the data. Remember to label your axes and to title your graph.

   b. Look at your graph and describe how this population of clams changed over time.

2. What factor limited the size of the clam population?

Part Two: Competition

3. a. Graph the population of clams and zebra mussels over time from Part Two of the Procedure. Use the same type of graph you used in Part One. Remember to label your axes and to title your graph. Use a key to show what represents the clam population and what represents the zebra mussel population.

   b. Look at your graph and describe how the population of clams changed over time.

   c. Look at your graph and describe how the population of zebra mussels changed over time.

4. a. What happened to the clam population after zebra mussels were introduced?

   b. Why did zebra mussels have this effect on the clam population? Explain.

5. a. In a real lake, what non-living factors might affect the size of clam and zebra mussel populations? List them. **Hint:** Go outside and look at an ecosystem around you. Observing an actual ecosystem may help you think of more factors.

   b. In a real lake, what living factors might affect the size of clam and zebra mussel populations? List them.

In this unit, you've learned to interpret population graphs and to analyze effects of factors such as competition, predators, and various environmental conditions on population size. Can a population graph tell you how much room there is for a particular species in a habitat? What does it mean for a population to run out of space?

What is carrying capacity?

READING

Imagine that you are a field ecologist. You've been studying a small lake called Lake Ness for the past ten years. You first began work at the lake when you heard that zebra mussels had invaded a nearby river, one that connects to Lake Ness. After ten years of study, you feel satisfied that you have a good idea of how quickly the zebra mussel can populate a lake of this size. You've been keeping an ongoing count of the zebra mussels in the lake (in mussels per square meter). At this point, your graph of population size looks like Graph 1.

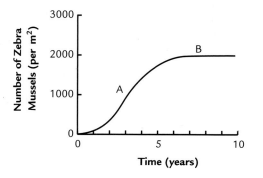

Graph 1: Zebra Mussel Population of Lake Ness Over 10 Years

STOPPING TO THINK 1

Recall that zebra mussels get their food by filtering plankton out of the water. Look at Graph 1. What do you think is happening to the quantity of plankton at:

a. Point A? Explain your reasoning.

b. Point B? Explain your reasoning.

As a result of your analysis, you think you have identified the maximum number of zebra mussels that could live successfully in Lake Ness. You think this might be the **carrying capacity** of the lake for zebra mussels. This term suggests the amount a container can hold, or carry. But unlike the capacity of a container, the number of zebra mussels that the lake can successfully hold may change over time, based on both living and non-living factors.

A few days later, your friend Nadia comes to visit you from the city. She drove up to the lake in her new car. "It has a carrying capacity of five passengers," she brags. Since you've never seen her drive anyone but her best friend and her dog, you simply shrug.

....

STOPPING TO THINK 2

a. Look again at Graph 1. What is the carrying capacity of zebra mussels in Lake Ness? How did you determine this?

b. List some of the factors that might affect this carrying capacity.

....

After Nadia leaves, you spend a week organizing your data. You decide to stop studying Lake Ness so closely for a while. Instead, you decide you'll return once a year to camp at the lake. During each visit, you can check on the zebra mussel population. Fifteen years pass. A graph of your data now looks like Graph 2.

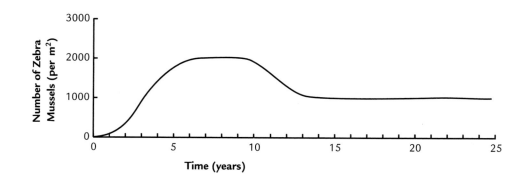

Graph 2: Zebra Mussel Population of Lake Ness Over 25 Years

STOPPING TO THINK 3

a. What is the carrying capacity for zebra mussels in Lake Ness between Years 13 and 25?

b. Identify at least three non-living factors that may have caused the carrying capacity to change. Explain how each factor could cause this change in carrying capacity.

c. Identify at least three living factors that may have caused the carrying capacity to change. Explain how each factor could cause this change in carrying capacity.

d. Do you think that the zebra mussel population will return to the level it had reached between Years 5 and 10? Why or why not?

For twelve years now, you've been puzzled by the change in the zebra mussel population. For example, in all your years of study, you've found no evidence of a new predator of zebra mussels appearing in the lake. You remain convinced that something about the zebra mussel's habitat must have changed to cause this shift in the population level. Consulting public records, you discover that a new factory was built just three miles from the lake about fifteen years ago!

Energized, you decide to test your hypothesis. You set up two identical tanks. One tank contains water from Lake Ness. The other tank contains water from a similar lake that is higher up in the mountains and farther from the factory. You add exactly ten adult zebra mussels to each tank. Every day, you supply the two tanks with fresh plankton, which you culture in a separate tank. Several months later, you are puzzled to find no difference at all in the zebra mussel populations of the two tanks.

STOPPING TO THINK 4

Is this a good experiment to test the hypothesis that the factory was affecting the zebra mussel population? Explain.

ANALYSIS

1. Shown below is the population graph from the Analysis section of Activity 78, "Coughing Up Clues."

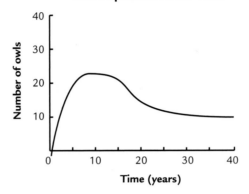

Owl Population Over Time

a. What is the carrying capacity for owls in this habitat?

b. How did the carrying capacity change during this 40-year period? Explain.

c. What living and non-living factors might explain this change in carrying capacity?

2. Turn back to Activity 72, "The Miracle Fish?" and look at Figure 2. Can you determine the carrying capacity of Nile perch in Lake Victoria based on this graph? Explain.

3. **Reflection:** Consider the introduced species you have been researching. Identify one ecosystem into which it has been introduced. Do you think this species has reached its carrying capacity in this ecosystem? Explain.

FIELD STUDY

Until now, you have focused on studying ecology in the laboratory. But ecology is the study of living organisms in the physical environment. This means that a majority of ecological study is done in the natural habitat of organisms, which is usually outdoors.

This type of outdoor investigation is known as **field study.** The scientist you read about in Activity 85, "Is There Room for One More?" performed a long-term field study of Lake Ness.

The "field" in field study can refer to any kind of ecosystem.

CHALLENGE

What do you observe when you conduct a field study?

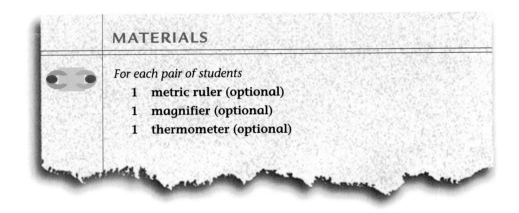

MATERIALS

For each pair of students
1 metric ruler (optional)
1 magnifier (optional)
1 thermometer (optional)

PROCEDURE

1. Select an ecosystem on your school grounds or near your school. Think about locations where you are most likely to observe interactions between living and non-living factors. Be sure to consider all of the possible habitats that are available in the area. For example, an overhanging roof may be home to a population of birds. Long grasses may contain many small animals, such as insects. Streams or ponds are also excellent places for field study.

2. Spend some time carefully observing your ecosystem. Start by simply sitting quietly and watching. Then record all the different types of habitat found within your ecosystem. For example, if you chose a small pond, you might identify the pond edge, the shallow water, and the deep water as three different types of habitat.

3. Record the characteristics of each habitat found within your ecosystem. For example, how much light and oxygen are available? How much rainfall is your habitat likely to receive? What is the temperature within the habitat? Will the temperature change a lot over a 24-hour period? Over the entire year?

4. Look for the presence of living organisms within your ecosystem. You may observe living creatures by gently looking among the different habitats, such as under leaves and rocks, or you may observe signs, such as animal tracks or other disturbances, that show that living creatures have been through the habitat.

5. Study your habitat for the next few days:

    a. Every day, observe your ecosystem for at least five minutes. Note any changes that occur. You may want to consider making your observations as an answer to a question, such as "Do I observe more species in the shady part of this ecosystem compared to the sunlit areas?"

    b. Quantitatively investigate one physical factor, such as temperature. Do this by taking measurements of this factor each time you observe your ecosystem.

6. If possible, create a food web for the organisms within your ecosystem. Identify the role (producer, consumer, or decomposer) that each organism plays within the ecosystem.

ANALYSIS

1. Summarize the results of your field study. What did you learn about this ecosystem? How did the physical factor you measured change over time? Was there any relationship between your observations and the physical factor you measured?

2. Compare the advantages of field study vs. laboratory work in studying ecology. Explain your ideas.

3. You may have seen documentaries or read books on ecosystems around the world. How do you think the information presented in these sources is gathered?

4. Many ecologists spend their entire lives studying a single ecosystem or population of organisms. For example, Dian Fossey spent almost 19 years studying the mountain gorillas of central Africa. Jane Goodall spent many years studying chimpanzees in their natural environment. Today, ecologists study ecosystems and organisms in all different parts of the world. Why do you think people spend their lives studying such systems? What can such studies tell us about the natural environment?

5. Reflection: How did field study differ from your laboratory work on ecology?

TALKING IT OVER

Having completed his research project, Ondar has a dilemma. He wants to do something about the problem of introduced species. He's particularly concerned about zebra mussels, which have been found in rivers and lakes around his state. What, if anything, should he do?

CHALLENGE

What are the trade-offs of trying to control an introduced species?

In this photo, you can see zebra mussel shells piled along the beach in a stack more than a foot high.

PROCEDURE

Read the statements of each of the following people. Decide what you would do if you were Ondar.

Johnson Poole, Engineer, Mantee Water Treatment Plant

"Zebra mussels cause a lot of problems for us. We supply water to the city of Mantee. It's our job to provide clean water for homes and businesses. To do that, large pipes bring water into the plant from Bear Lake. Here at the plant, we filter and treat the water before sending it on to the city.

"But we've had a hard time lately getting the water into the plant. Those zebra mussels grow on everything, including the insides of the pipes. We have seen up to 750,000 zebra mussels in a square meter of pipe! As you might imagine, all of these zebra mussels begin to block the flow of water.

A worker uses a hose to suction zebra mussels from inside a pipe at a power plant.

"Right now, we shut down the plant every few months. Then we send someone into the pipes to physically remove all the mussels. This costs tons of money—the U.S. Fish and Wildlife Service reports that dealing with this problem in the Great Lakes area alone has cost billions of dollars!

"In the meantime, we're looking at other solutions. For instance, we're exploring ways to prevent zebra mussels from settling and growing on the pipes in the first place. Zebra mussels grow best on hard surfaces, such as rocks. That's also why they sometimes grow on other animals with hard shells like clams. We're trying to find out if we can coat the pipes with some type of paint or something else that would prevent the mussels from growing on them. You could say we're trying to make the pipes a less suitable habitat for the mussels!"

➤

Adrienne Vogel, Chemist, Bear Industrial Company

"Our company uses water from Bear Lake. Chemicals have been shown to kill both larval and adult zebra mussels. That's one way we prevent zebra mussels from growing in our water supply. We can't afford for zebra mussels to grow inside our water containment ponds. So, after the water comes into our plant, we treat it with a variety of chemicals. While this is very effective in dealing with the zebra mussels, the treated water does contain a lot of chemicals. This means that we can't release the water back into the lake as is. Luckily for us, we are able to recycle and re-use the water within the company for several months. Before we release the water, we treat it to isolate the chemicals and dispose of them according to state regulations. But this all costs money."

Talia Mercata, Biologist, State Fish and Wildlife Service

"I sympathize with the people at both plants. Humans are not the only ones that are affected by zebra mussels. Zebra mussels may be changing the native ecology of lakes and rivers. We know that Bear Lake is clearer as a result of zebra mussels. Some people think this is a good thing. One thing is certain—zebra mussels have filtered out so much of the algae that they may be changing the feeding structure of the lake.

"Some scientists are investigating how predators may help control zebra mussel populations. In Europe, where zebra mussels first came from, there are a lot more native predators, such as fish that have teeth. Here in the U.S., the populations of fish that might be good predators aren't that high.

"Ducks are one possible predator here in the U.S. But using predators to control zebra mussels is complicated. How do you control where ducks and fish decide to search for food? How can you guarantee that they'll eat zebra mussels and not some other food? How will they reduce populations in hard-to-reach areas, such as inside pipes? What happens if the introduction of the predator causes other imbalances in the ecosystem?

"Because of these difficulties, my research focuses on the use of parasites to control zebra mussels. If my research is successful, I may identify a parasite that could infect and kill zebra mussels. I'm not sure how quickly this would affect their populations, though."

Henry Wai, Activist, People for Responsible Action

"It's a shame that zebra mussels were ever introduced into the United States. We can only predict how they'll affect the ecology of our lakes and rivers. We don't know for sure.

This outboard motor is covered with zebra mussels.

"People caused this problem in the first place and I think every person should take responsibility for trying to prevent further damage. It's easy to forget that things we do every day might contribute to the problem of introduced species, but it's true.

"For example, just carrying equipment like inner tubes and diving gear from one lake to another can introduce a species like the zebra mussel. After all, its larval stage is very small. That's why it's important for people to rinse and dry their equipment before going from one body of water to another. Think about it—if every boater, fisher, swimmer, and diver took care to clean off his or her equipment, we might prevent zebra mussels and other organisms from spreading across the U.S. so quickly!"

EXTENSION

Go to the SALI page of the SEPUP website for links to sites with information about zebra mussels and management options.

ANALYSIS

1. What, if anything, do you think should be done about the growing population of zebra mussels in the United States? Support your answer with evidence and discuss the trade-offs of your decision.

Hint: To write a complete answer, first state your opinion. Provide two or more pieces of evidence that support your opinion. Then discuss the trade-offs of your decision.

Introduced species can have an enormous impact on the economy as well as on native ecosystems. Your research project and your study of ecology have helped you to become an expert on one introduced species. Why are some introduced species more likely to be successful than others?

CHALLENGE

What, if anything, should be done about the introduction of a new species into an ecosystem?

These workers are removing hydrilla and other aquatic plants from a lake.

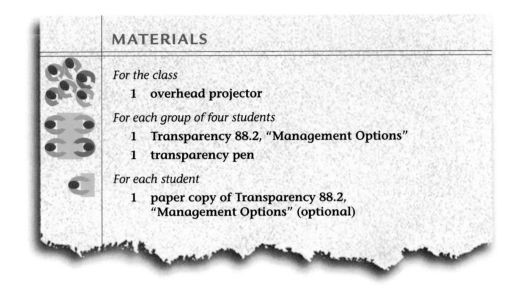

MATERIALS

For the class
   1   overhead projector

For each group of four students
   1   Transparency 88.2, "Management Options"
   1   transparency pen

For each student
   1   paper copy of Transparency 88.2,
       "Management Options" (optional)

PROCEDURE

1. In Activity 73, "Introduced Species," you began a research project on an introduced species. You will now present your research to the class. Use Student Sheet 73.1, "Introduced Species Research," as you plan your presentation. Your presentation should help your audience make an informed decision about what, if anything, to do about this introduced species.

   When planning your presentation, remember:

   • All the members of your group must participate.

   • Since any group member may be asked to answer questions from the class, all group members should fully understand the report.

   • Your presentation time is limited.

   • Many people learn best from a mix of visual, written, and spoken information. Include graphs and maps when possible.

   • While you have your own opinions on this issue, it is important that you present unbiased and complete information. The members of your audience can then make their own decisions.

   • You may want to role-play different experts when presenting your information, such as the people who might present information at a city council meeting. The class would represent the community members who would be voting on a decision.

2. List all of the options that are available for dealing with your introduced species on Transparency 88.2, "Management Options."

3. Begin by presenting general information about your introduced species to the class. Respond to any questions that other students may have.

4. Ask the class what they think are the pros and cons of each of the options you presented. Record their responses on Transparency 88.2.

5. If you are aware of issues that were not brought up by the class, add them onto the transparency.

6. Have the class vote on what, if anything, should be done about the introduction of this species into new ecosystems.

7. Listen to and participate in other groups' presentations.

ANALYSIS

1. Many species are accidentally introduced into North American ecosystems from other countries each year. The opposite is also true: North American species are also introduced into other countries.

   a. What other countries or other areas of the United States are most likely to exchange species with the area where you live?

   b. Only a small fraction of species that are introduced are successful enough to create problems in their new environment. What features of a species do you think make it likely to be successful in a new environment? Use specific examples from the project presentations in your answer.

2. How do you think the number of introduced species in the United States will change over the next 50 years? Explain your reasoning.

3. Write a letter to the editor of a local newspaper describing the situation of an introduced species. Explain what, if anything, you think should be done about the species. Support your answer with evidence and discuss the trade-offs of your decision.

   Hint: To write a complete answer, first state your opinion. Provide two or more pieces of evidence that support your opinion. Then discuss the trade-offs of your decision.

Evolution

F

Unit F

Evolution

It was Kenya's fourth visit to the pet store. Ever since she decided she wanted a pet lizard for her birthday, she had tried to come every day. She still hadn't decided which lizard she would like to have—and her birthday was less than a week away!

"Excuse me, young lady, can I help you?" asked the sales clerk behind the counter.

"I want a lizard for my birthday," replied Kenya. "But I can't decide which one I like best. There are so many different kinds—and they look so different."

"Some of them eat different foods, too," added the sales clerk.

"I don't understand how there can be so many different kinds of the same animal," said Kenya. "It's amazing! I wonder how it happened."

• • •

Have you ever wondered about the amazing variety of organisms on Earth? How did they evolve? How are they related? Just as historians study the history of humans, some scientists study the history of life on Earth. They do this by gathering evidence, making connections, creating models, and testing theories. In this unit, you will learn to interpret the many sources of evidence that exist for the evolution of life on Earth.

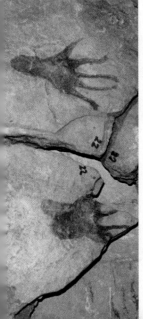

When the last member of a species dies without any surviving offspring, we say that that species has become **extinct**. Every species alive today is related to many other species that have already become extinct. Becoming extinct is not a sign of inferiority, but just another sign that ecosystems are constantly changing. In fact, it is estimated that 99.9% of all species that have ever lived on Earth are now extinct. Today, species that have such a small population that they are in danger of becoming extinct are called **endangered species**.

CHALLENGE

What are the trade-offs in deciding whether to save an endangered species or to re-create an extinct one?

Extinct animals include dinosaurs and saber-toothed cats.

Endangered animals include tigers and gorillas.

PROCEDURE

Work with your group to read and discuss the story of mammoths and modern elephants.

Mammoths and Elephants

You may know that dinosaurs became extinct about 65 million years ago, 64 million years before humans evolved. There is evidence that at least one enormous asteroid crashed into Earth at that time. Many scientists believe that this created huge clouds of dust that blocked out the sun for a long period of time. As you learned in the Ecology unit, producers form the base of the food web. A loss of sunlight would cause the death of many producer species, which, in turn, would cause the death of many consumer species, such as dinosaurs. By the time the dust settled and sunlight could reach Earth's surface, thousands of species, including the dinosaurs, had become extinct and most ecosystems were greatly changed.

One species that became extinct much more recently is the mammoth. If mammoths were still around, they would be close relatives of the elephants living on Earth today. The entire bodies of some mammoths were trapped during the most recent ice age and have remained frozen ever since. Explorers have tasted mammoth meat, as have several curious scientists! Some scientists think that the tissue of frozen mammoths is in good enough shape to bring mammoths back from the dead.

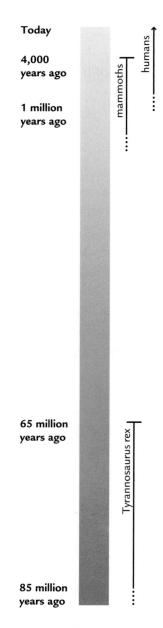

Today

4,000
years ago

1 million
years ago

65 million
years ago

85 million
years ago

Figure 1: Timeline

Mammoths evolved 3 to 4 million years ago, about 60 million years after dinosaurs became extinct (Figure 1). Mammoths thrived and spread to North America about 1.8 million years ago. But about 10,000 years ago, all but a few small herds of mammoths died. The last mammoth died around 4,000 years ago. There is no evidence that an asteroid or other catastrophic event brought about the extinction of the mammoths.

What did happen 10,000 years ago that caused this huge drop in the mammoth population? One possibility is that the mammoths could not survive the drastic changes in climate and vegetation that occurred when the last ice age ended. In addition, humans—who lived at the same time as the mammoths—were moving into new environments as their population grew. The end of the last ice age helped expand the range of humans into areas where mammoths lived. Increased hunting of mammoths by humans may have contributed to their extinction.

While mammoths and modern elephants are closely related, mammoths are not direct ancestors of modern elephants. In fact, until the mammoth became extinct, mammoths and elephants were alive in different parts of the world. Based on fossil remains, the common ancestor of both modern elephants and mammoths is estimated to have lived 4 to 5 million years ago. The fossil considered to be the first member of their order is dated at about 55 million years ago. Since then, scientists believe that there have been over 500 different elephant and mammoth species. Only two of these species are alive today: the Asian (or Indian) elephant and the African elephant. Figure 2 shows a "family tree" including modern elephants and several extinct relatives. Populations of both African and Asian elephants are declining, and the Asian elephant is considered an endangered species.

The Asian elephant is smaller than the African, with smaller ears and a slightly rounded or flat back. Asian elephants have a double-domed forehead (African elephants have only a single dome). In addition, Asian elephants have a single "finger" on the upper tip of the trunk, while African elephants have a second on the lower tip.

African elephant   *Asian elephant*

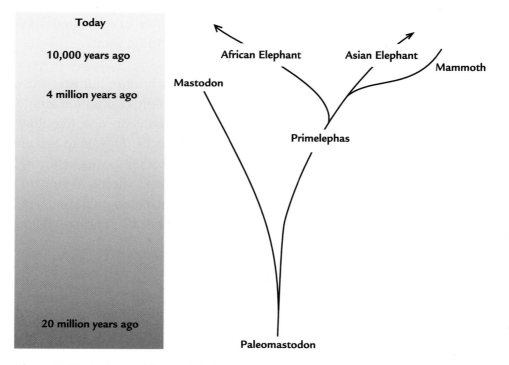

Figure 2: Evolution of Modern Elephants

Unlike African elephants, which all have tusks, only the male Asian elephants have them. In fact, even some of the male Asian elephants do not grow tusks! Killing elephants for their ivory is illegal in India and China. Still, most of the large-tusked male Asian elephants have already been killed for their ivory.

Asian elephants used to roam from Iran to southern Asia (Figure 3). In the early 1900s, approximately 250,000 Asian elephants lived in the wild. Today, it is estimated that no more than 50,000 Asian elephants are left. Their population has declined by more than 80% in less than 100 years! Without intervention, Asian elephants will most likely become extinct. By passing laws, raising money, creating wildlife preserves, and raising awareness, some people are working hard to save the Asian elephant.

Not all people are fighting to save Asian elephants. Asian elephants are forest animals. As the human population increases, forests have been cut down to build farms and villages. Today, most wild Asian elephants have been forced to live in hill and mountain regions. A single adult elephant eats about 330 pounds of grasses, roots, leaves, and bark each day, and these environments cannot always supply enough food. Elephant herds often seek out nearby farms that grow crops such as sugar cane and grains. These farms suffer crop loss, property damage, and even loss of life. In an average year, Asian elephants kill approximately 300 people in India alone.

Loss of habitat, combined with human hunting, has caused the decline in the Asian elephant population, a situation similar to that faced by the mammoth several thousand years ago. Should the Asian elephant be saved, or should this species be allowed to become extinct, just like the mammoth and millions of other species before it? Are people spending too much time, energy, and money trying to save endangered species? Or should efforts be increased, perhaps by going so far as to try to re-create extinct species, as has been proposed for the mammoth?

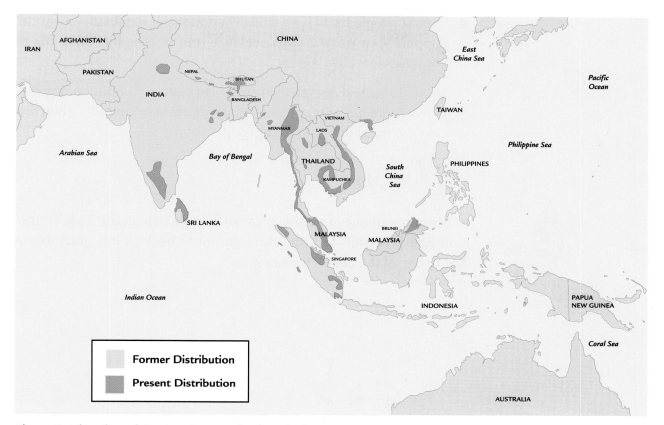

Figure 3: Historic and Current Range of Asian Elephants

ANALYSIS

1. What are the similarities and differences between the extinction of mammoths and the possible extinction of Asian elephants?

2. Use evidence from this activity to explain why the mammoth could once have been considered an endangered species.

3. Some scientists would like to try to re-create a living mammoth by removing the DNA from a fertilized elephant egg and replacing it with mammoth DNA.

   a. Which species of elephant egg do you think scientists should try first? **Hint:** Look carefully at Figure 2.

   b. Do you think scientists should try to re-create a living mammoth? Explain.

4. Should people try to save wild populations of the Asian elephant? Support your answer with evidence and discuss the trade-offs of your decision.

Hint: To write a complete answer, first state your opinion. Provide two or more pieces of evidence that support your opinion. Then discuss the trade-offs of your decision.

EXTENSION

Learn more about attempts to save the Asian elephant from extinction and proposals to bring the mammoth back to life. Start at the SALI page of the SEPUP website.

Many species have become extinct during the history of Earth. How can you know these creatures ever existed? The evidence is right under your nose—or your feet, to be more precise.

Our planet's thin outer layer, the crust, can be up to 40 kilometers (25 miles) thick. The crust is made up of many layers of rock that have been forming for over 4 billion years, and are still forming today. These rock layers can form when a volcanic eruption covers the land with lava, or when a flood spreads out a layer of mud. Lava, mud, or even sand can eventually harden into solid rock. New rock layers can also form over hundreds of years as sediment—sand, dirt, and the remains of dead organisms—gradually settles on the bottom of a lake or ocean.

Any new layer of rock can seal off the layer below it. Organisms trapped within these sealed off layers can become part of the rock itself. Any trace of life preserved in a rock is called a **fossil.** It can be an entire organism, a part of an organism, a footprint, a piece of feces, or a piece of shell, bone, or tooth.

CHALLENGE

What can fossils tell you about organisms that lived in the past?

MATERIALS

For the class
8   containers of fossils

For each pair of students
1   hand lens

For each student
1   Student Sheet 90.1, "Fossil Observations"

PROCEDURE

1. Work in a group of four. Collect a pair of fossils. One pair in the group should begin by examining one of the fossil specimens, while the other pair begins by examining the other specimen.

2. Work with your partner to identify the unique features of the fossil. Be sure to look at both specimens of the fossil species. Use the magnifier to help you.

3. On Student Sheet 90.1, "Fossil Observations," sketch the general shape and unique features of this type of fossil. Then record additional observations that are difficult to show in your sketch, such as color or size. Note that your group of four has two specimens of the same fossil. You can write observations on both of these specimens.

4. When directed by your teacher, exchange your pair of fossils with another group of four students.

5. Repeat Steps 1 through 4 until you have examined all eight types of fossils. As you continue to look at more fossils, observe similarities and differences among the different fossils.

ANALYSIS

1. Review your notes on the eight different types of fossils. Do you think any of them are from similar species? Explain, using evidence from this activity to support your answer.

2. In this activity, you were given a fossil to examine. What additional observations could you have made about the fossil if you had discovered it yourself?

3. Choose one of the eight fossils you examined.

   a. Based on the fossil, describe what you think this organism looked like when it was alive. Include your evidence for your description.

   b. In what type of environment would you expect to find this organism? Explain your reasoning.

4. Although you probably have a vivid picture of dinosaurs in your mind, no one has ever seen a living dinosaur. All the evidence for the existence of dinosaurs comes from fossils.

   a. What details about the appearance and behavior of dinosaurs do you think would be easiest to determine from fossils?

   b. What details about the appearance and behavior of dinosaurs do you think would be hardest to determine from fossils?

Paleontologists (pay-lee-uhn-TALL-uh-jists) are scientists who study fossils. Fossils are rarely complete and are often just a shell, half a leaf, or a couple of bones. In some cases, the only evidence left by an organism is its tracks. Footprints and other types of animal tracks can be fossilized in the same way as actual body parts. But what can you find out from just footprints? Like detectives, paleontologists can use the information from fossil footprints to determine how an organism moved, how fast it traveled, what type of environment it lived in, and what it might have been doing when its footprint was formed.

Few fossil remains are as complete as this 10 million-year-old rhinoceros in Nebraska.

Fossil footprints

EVIDENCE COMES IN STEPS

A fossil footprint site has just been discovered! You take a helicopter to the location in the hope that your expertise will be useful. The rest of the team is slowly brushing away layers of sediment to carefully uncover the footprints.

Your task is to use your observations to draw inferences and then develop a hypothesis about what happened to form the footprints. As the footprints are uncovered, there will be more evidence to examine. Remain open to new possibilities as the investigation continues.

CHALLENGE

How can fossil footprints be used to study the behavior of animals that were alive millions of years ago?

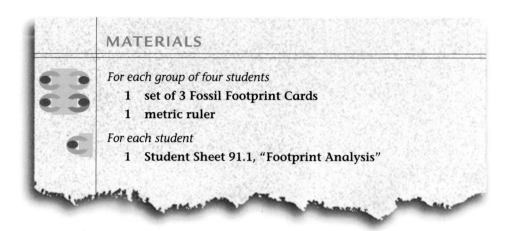

MATERIALS

For each group of four students

1   set of 3 Fossil Footprint Cards
1   metric ruler

For each student

1   Student Sheet 91.1, "Footprint Analysis"

PROCEDURE

1. Examine Fossil Footprint Card 1, which shows what the team has uncovered so far.

2. In your group, discuss what you think was happening while these footprints were being created. You do not have to agree, but:

   - If you disagree with others in your group about what happened, explain to the rest of the group why you disagree.

   - Listen to and consider other people's explanations and ideas.

3. Record your ideas in the first row of Student Sheet 91.1, "Footprint Analysis." Separate your ideas into observations and inferences. **Note:** Even though some of your inferences may conflict with other inferences, consider as many ideas as possible.

4. Time passes and more footprints are uncovered. Obtain Fossil Footprint Card 2.

5. Repeat Step 2. Then record your additional observations and inferences in the second row of Student Sheet 91.1. However, do not change what you wrote in the first row!

6. Time passes and a third section of footprints is uncovered. Obtain Fossil Footprint Card 3.

7. Repeat Step 2. Then record your additional observations and inferences in the third row of Student Sheet 91.1. Remember, do not change what you wrote in the first two rows!

8. Look back at all your observations and inferences. Try to think of the best possible explanation for how the footprints were formed. Record your strongest hypothesis in your science notebook. If you have two or more hypotheses in mind, record them all, but rank them from most likely to least likely.

ANALYSIS

1. Why is it important for scientists—and people in general—to distinguish between observations and inferences when they develop a hypothesis?

2. Imagine that the team uncovered a fourth section of footprints. Draw what you predict this fourth section might look like. Explain how it would provide more support for the hypothesis you favor.

3. Different types of information can be collected from footprints. In addition to observing the shape, size, and arrangements of footprints, their depths can be measured. The tables below show two different sets of measurements that might have been taken.

Table 1 Average Depths of Footprints (Scenario 1)			
	Card 1	**Card 2**	**Card 3**
Larger footprints	6.0 cm	6.2 cm	8.3 cm
Smaller footprints	2.5 cm	2.6 cm	———

Table 2 Average Depths of Footprints (Scenario 2)			
	Card 1	**Card 2**	**Card 3**
Larger footprints	6.0 cm	6.2 cm	6.1 cm
Smaller footprints	2.5 cm	2.6 cm	———

   a. What hypotheses would the data in Table 1 support? Explain how these data would provide more evidence in support of one or more hypotheses.

   b. What hypotheses would the data in Table 2 support? Explain how these data would provide more evidence in support of one or more hypotheses.

   c. What factor(s) might explain the difference in the depth of the footprints in the different sections?

➤

4. a. Think back to an activity in which you came up with hypotheses based upon evidence, such as Activity 74, "Observing Organisms," in the Ecology unit. Describe an example of an observation and an inference based upon that observation and explain how the two are different.

   b. Describe an example of an observation and an inference from a recent event in your everyday life.

MODELING

As you learned in Activity 90, "Figuring Out Fossils," the history of Earth is divided into time spans. These time spans do not last any specific number of years. The beginnings and endings of the time spans are determined by fossils—either the appearance of new types of fossils that are not found in any older rocks or the disappearance of fossils that are commonly found in older rocks. With the help of radioactive dating technology, scientists have made good estimates of how many years each time span lasted.

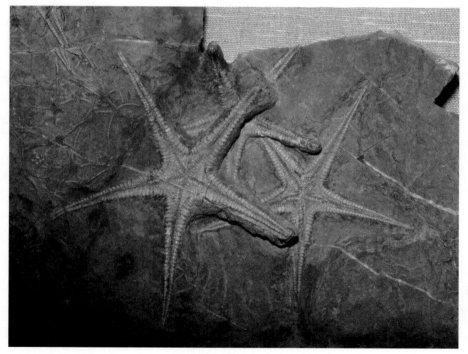

Jurassic sea star fossils

CHALLENGE

How long have organisms been living on Earth?

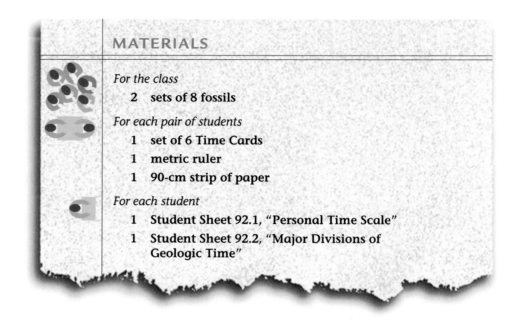

MATERIALS

For the class
  2   sets of 8 fossils

For each pair of students
  1   set of 6 Time Cards
  1   metric ruler
  1   90-cm strip of paper

For each student
  1   Student Sheet 92.1, "Personal Time Scale"
  1   Student Sheet 92.2, "Major Divisions of Geologic Time"

PROCEDURE

Part One: A Personal Time Scale, Geologic-Style

1. Look at the following list of events. Write the event that occurred most recently at the top of the column labeled "Order of Events" on Student Sheet 92.1, "Personal Time Scale."

   I started fourth grade.

   I ate or drank something.

   I learned to walk.

   I woke up.

   I was born.

   I took a breath.

   I started kindergarten.

   I learned to read.

   My parents were born.

2. Use the remaining spaces in the "Order of Events" column to write down the other events from most recent (at the top) to most distant (at the bottom).

3. In the column "Number of Years Ago," write the number of years ago that each event occurred (you can round off to the nearest year, or half-year). Like a paleontologist, count time backward from the present day. For example, if the event occurred 10 years ago, write "10 ya" as the time of the event. (The unit "ya" means "years ago.")

These students were born about 14 ya.

4. Think of a major event in your life that is important to you. (It may or may not already be described in your "Order of Events" column). Use this event to divide your time scale into two time periods by drawing a horizontal line to mark when the event occurred. For example, if you choose entering school as the major event, you could draw a line right below "I started kindergarten."

5. Name the two time periods that you just created. For example, if you drew a line at the time you first started school, the time period before that could be called "Pre-Schoolian."

6. As a class, compare the events that you and your classmates chose to divide your personal time scales into two periods. Work together to agree on a single event that was important to everyone in class. Agree on names for the time periods before and after that event.

➢

Part Two: Geologic Time

7. Imagine that a paleontologist asks you to help her put in order some periods of time in the history of life. With your partner, read carefully the information on the six Time Cards and arrange them with the oldest on the left and the most recent on the right.

8. In your science notebook, record the order in which you placed the cards.

Mount Rainier formed approximately one million ya (1 mya).

9. View the work of other student groups. Observe the similarities and differences between their orderings and yours. Discuss why you made the choices that you did.

10. Obtain Student Sheet 92.2, "Major Divisions of Geologic Time," and a 90-cm strip of paper from your teacher. Use the information on Student Sheet 92.2 to arrange the cards in the order scientists have determined from geologic evidence. In your science notebook, record any changes that you needed to make to your original order.

11. Follow Steps 11a–d to construct a timeline of the last 4,500 million years:

   a. Using Student Sheet 92.2, work with your partner to calculate the distance (in cm) that each time span will cover on your timeline.

   Hint: Since your timeline must represent 4,500 million years over 90 centimeters, first divide 4,500 by 90 to determine how much time each centimeter will represent.

   b. Draw a vertical line near one end of your long strip of paper and label it "The Origin of Earth."

   c. Using "The Origin of Earth" as a starting line, use a ruler and your calculations from Student Sheet 92.2 to mark the boundaries between the time spans.

   d. Label each time span with its name and each boundary with its defining event.

12. Figure 1 on the next page presents photos of the fossils you examined and sketched in Activity 90, "Figuring Out Fossils." In the appropriate time period on your timeline, draw and label a quick sketch or outline of each one.

Figure 1: A Few Familiar Fossils

Admetopis subfusiformis

Knightia

Alethopteris serii

Ammonite

Goniobasis tenera

Tabulopyllum

Mucrospirifer thedfordensis

Elrathi kingi

EXTENSION 1

Obtain a copy of a more detailed geologic time scale. Construct a timeline that represents only the last 550 million years. Label all the *periods* with their names and be sure to distinguish them from the *eras*. What additional information were you able to include on this time-line? What are the advantages and disadvantages of creating time-lines for shorter time periods?

EXTENSION 2

As a class, create a giant timeline that represents some of the major events (such as the first fossils of interesting life forms, mass extinctions, etc.) that have occurred during the 4.5 billion-year history of Earth. Stand at appropriately scaled distances from your classmates, and together hold up signs representing major events in the history of life.

ANALYSIS

1. Think back to how you and your classmates divided your personal time scales into periods. How do you think scientists determined how to divide geologic time into its periods?

2. The total length of your timeline of Earth's history is 90 cm. Use your timeline to determine the fraction of Earth's history that:

   a. single-celled organisms have lived on our planet.

   b. multicellular organisms have lived on our planet.

3. **Reflection:** Imagine that no species ever became extinct. Do you think there would be more, less, or the same amount of diversity of life forms on our planet? Explain your answer.

INVESTIGATION

In some places, such as the walls of a deep river canyon, hundreds of rock layers are visible, one on top of the other. As rock layers form, each new layer is deposited on top of an already existing layer. When you observe a sequence of rock layers, the top layer, along with any fossils it contains, is younger than any other layer in that sequence, and the bottom layer, along with any fossils it contains, is the oldest layer in that sequence. This is called the **law of superposition.**

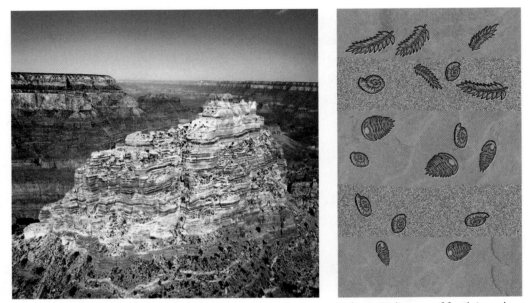

Rock layers in the Grand Canyon

Schematic diagram of fossils in rock layers

A diagram representing a series of rock layers, such as the one on the right, is called a **stratigraphic column.** Stratigraphic columns can be made by looking at the sides of cliffs, or by looking at drill cores. A drill core is a cylindrical piece of rock removed from the Earth by a large drill, similar to the drills that are used to make oil wells. Drill cores can provide samples from many miles beneath the surface of the Earth.

No single location contains a complete set of all the rock layers or fossils that exist on Earth. In order to study a particular fossil organism or find out which organisms lived during which geologic era, paleontologists must compare rocks from different places throughout the world. You will examine and compare four different drill cores, each representing the rock layers found on different fictitious continents.

CHALLENGE

How can you determine which fossils are older, which are younger, and which are likely to be from extinct species?

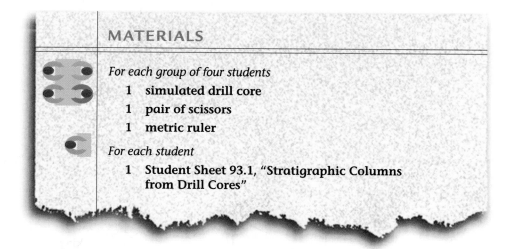

MATERIALS

For each group of four students

1 simulated drill core
1 pair of scissors
1 metric ruler

For each student

1 Student Sheet 93.1, "Stratigraphic Columns from Drill Cores"

PROCEDURE

1. Examine your drill core. The top of each drill core is marked with its number.

2. Create a stratigraphic column by sketching in the boundaries of the layers and the fossils found within each layer in the appropriate place on Student Sheet 93.1, "Stratigraphic Columns from Drill Cores."

3. Based on the evidence within the layers of this drill core, list the fossils in order from youngest to oldest.

4. When directed by your teacher, exchange your drill core with a group that has a drill core with a different number.

5. Repeat Steps 1–4 until you have observed, sketched, and analyzed all four drill cores.

6. Based on the appearance of the rock layers and the fossils found within each layer, match, or correlate, the layers from each core as best you can. Make a chart, similar to the one in Figure 1, that shows your correlation of the rock layers from the four different drill cores.

Hint 1: You may want to cut out each column from the Student Sheet so that you can move them around as you try to match up the layers.

Hint 2: Layers don't have to be exactly the same to correlate.

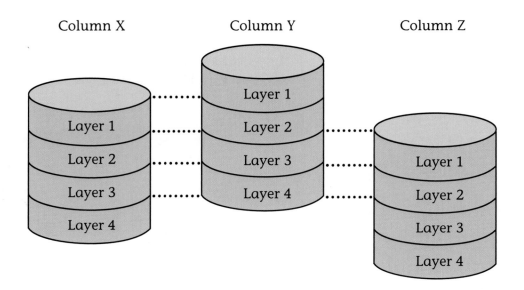

Figure 1: Sample Correlation of Stratigraphic Columns

7. Use your correlation chart to list all four of the fossils in order from youngest to oldest.

Hint: If you think a layer found in one drill core is the same as a layer found in another drill core, you can infer that those layers, and the fossils in them, are the same age.

ANALYSIS

1. Describe some of the difficulties you had trying to match evidence found in one drill core with evidence found in other drill cores. What additional evidence would have helped you make your correlations?

2. Based on evidence from all four drill cores, which, if any, of the organisms represented by the fossils may be from species now extinct? Explain.

3. Which fossil species could have lived at the same time?

4. Using the information below and the list you made in Step 7 of the Procedure, make a timeline that shows the time span when each species is believed to have been alive. Explain how you determined your answer and whether it is based on inference, observation, or a combination of both.

Core	Layer	Geologic Era
4	1	Early Cenozoic
1	2	Early Mesozoic
3	5	Middle Paleozoic
2	5	Early Paleozoic

Hint: Refer to Figure 1 in Activity 89, "Mammoth Mysteries," on page F-6 for help in designing your timeline.

5. **Reflection:** Propose what might have caused the changes through time shown on your timeline. Explain.

ROLE PLAY

Plenty of fossil evidence shows that most of the species that have lived in the past are no longer alive today. It also seems that most of the species on Earth today were not always here. In other words, different species of organisms have lived at different times in Earth's history. New species have descended from earlier species, but have changed over long periods of time. These changes through time are called **evolution.**

But how does evolution happen? Two major theories were proposed during the 19th century. The first was disproved and abandoned, while the second has helped evolution become a central idea in modern biology. What would it sound like if the original experts met and discussed the problem?

CHALLENGE ➡

How does evolution happen?

PROCEDURE

1. Assign a role for each person in your group. Assuming there are four people in your group, each of you will read one role.

   Roles

   Charles Darwin, 19th century scientist

   Isabel Matos, science reporter for Station W-EVO

   Jean-Baptiste Lamarck, 19th century scientist

   Wendy Chin, middle school student

2. Read the role play on the next pages aloud. As you read, think about what each character is saying.

HOW DO SPECIES EVOLVE?

Isabel Matos: In today's episode of "Time Travel News," we have brought together two of the first scientists to publish ideas on how evolution occurs. Visiting us from the 19th century are Jean-Baptiste Lamarck and Charles Darwin. Monsieur Lamarck, let's start with you.

Jean-Baptiste Lamarck: I was one of the first to recognize that species evolve. In 1809, I proposed the first theory of how evolution occurs. Allow me to explain my theory. Let's begin by talking about giraffes. Wendy, why do you think giraffes have such long necks?

Wendy Chin: To reach leaves at the tops of trees, I guess. They have to be able to get food.

Lamarck: Quite right. I began to wonder how giraffes' necks became so long.

Wendy: I bet they evolved that way.

Lamarck: But how did this evolution occur?

This is what I wanted to understand. My theory was that giraffes stretched their necks by reaching for leaves that were higher and higher on the trees. This made their necks longer. Then, when they had babies, their babies had longer necks too. Look—this sketch helps explain my ideas.

| This is an adult giraffe. | The giraffe reaches for leaves slightly out of reach. | The use of the neck causes it to lengthen slightly. | The offspring of the giraffe also has a longer neck. |

Figure 1: Lamarckian Evolution

Wendy:	Shouldn't a theory be based on evidence?
Matos:	Mr. Lamarck, did you ever see an adult giraffe grow its neck longer?
Lamarck:	Of course not. My idea was that the growth was very small, too small to measure in one generation.
Charles Darwin:	I'd like to explain another theory, called natural selection. Alfred Russel Wallace and I constructed this theory at about the same time. We also noticed that not all animals of the same type have the same features. Take horses, for instance.

Wendy: Oh, I know what you mean! There are horses of different sizes and colors, but they are all one species and can interbreed.

Darwin: Exactly—and the same is true of giraffes. Have you noticed that animals in the same species look different, or varied? This is important because, in the wild, some animals in each species usually die every year. Only animals that survive can give birth to offspring. Now, what feature of a giraffe might help it to survive and live to reproduce?

What differences do you observe in these giraffes of the same species?

Lamarck: Its neck, of course! As I said before, it must stretch from being used so vigorously. Giraffes can then pass on the longer necks to their children.

Matos: But Mr. Lamarck, modern scientists have found no evidence for your hypothesis that parents can pass *acquired* traits to their offspring. Consider professional wrestlers. They build muscles by lifting weights. But their babies are no stronger than other babies. If these babies want to have muscles like their parents, they have to pump a lot of iron too!

Darwin: But just like human babies, not all giraffes are the same. They have slight differences in all their characteristics, including neck length.

Lamarck: So you're saying any giraffe that happens to have a slightly longer neck can eat leaves that are higher in a tree than a shorter-necked giraffe can and therefore is more likely to survive.

Wendy: So the longer-necked giraffes are more likely to live longer because they can reach more food. If more of these giraffes live longer, they can produce more offspring!

Darwin: That's right. Animals with certain features, such as giraffes with longer necks, are more likely to live to adulthood and have more babies. We call that process **natural selection**. Here's a sketch of how it works

Giraffes with longer necks tend to reach leaves more easily.

Longer-necked giraffes are more likely to eat enough to survive . . .

. . . and reproduce. The offspring inherit their parent's longer necks.

Figure 2: Darwinian Evolution (Natural Selection)

Wendy: But why will the offspring of longer-necked giraffes have longer necks too?

Matos: Well, tall parents are more likely to have tall children, aren't they? The same is probably true of giraffes.

Darwin: According to my theory, each new generation of giraffes has, on the average, slightly longer necks than the generation before.

Lamarck: But not because they stretched their necks? Only because the longer-necked giraffes were more likely to survive and reproduce?

Wendy: I get it. Individual animals don't change, but over very long periods of time, the population of an entire species does.

Lamarck: But, Mr. Darwin, can your theory of natural selection explain why extinction occurs?

Darwin: I believe so. Consider the mammoth, which became extinct a few thousand years ago. Why didn't mammoths evolve and continue to survive?

Wendy: There are several theories about that. They became extinct during a time when the global climate was warmer than it had been before. The changing climate may have affected the mammoth's food supply, and human hunters may have contributed to the extinction.

Matos: So a species becomes extinct when it doesn't survive an environmental change. No individuals in the population have the traits necessary to survive.

Darwin: That's all it is. The **variation** in the population isn't enough to withstand environmental changes. In fact, sooner or later, most species become extinct.

Wendy: Let me get this straight. As time passes, species change, and we call this evolution. The way this occurs is by natural selection—some individuals in a population happen to be better suited to the environment and they're more likely to survive and reproduce.

Lamarck: As a result, the population as a whole over many generations comes to have an **adaptation**, such as a giraffe's longer neck.

Matos: Today, we know that we pass on characteristics like longer necks to our offspring through genes. Genes don't change because you exercise your neck.

Darwin: Tell us more about these genes.

Wendy: I learned about genes in school. Genes are things in our cells that we inherit from our parents. They cause us to have traits—the way we look and stuff.

Lamarck: Fascinating. I would like to learn more about this.

Darwin: Without this modern evidence, I hesitated to publish my theory for years, until Wallace sent me a brief paper containing the same ideas. Within a few years of our publications, scientists widely accepted the idea that species arise by descent with modification, or evolution.

Matos: Thank you, Mr. Lamarck and Mr. Darwin. Viewers, I hope you've enjoyed meeting people from our past. Join us next week for a scintillating conversation with Marie Curie, the first woman scientist to receive a Nobel Prize.

➤

ANALYSIS

1. a. Compare and contrast Lamarck's and Darwin's theories of evolution: What are the similarities? What are the differences?

b. Why do scientists find Darwin's theory more convincing?

2. Ancestors of modern elephants had much shorter trunks than elephants do today. Use Lamarck's theory of evolution to explain how the trunks of elephants might get longer over many generations. Drawing a picture may help you to explain what you have learned.

3. Use the Darwin/Wallace theory of natural selection to explain how the trunks of elephants might get longer over many generations. Drawing a picture may help you to explain what you have learned.

4. Reflection: If you have completed Unit C, "Micro-Life," of *Science and Life Issues,* look back at Activity 51, "The Full Course," and Activity 52, "Miracle Drugs—Or Not?" to review antibiotic-resistant bacteria. How is the problem of antibiotic resistance in bacteria an example of natural selection?

MODELING

n the last activity, you considered the interaction between the environment and a species over a long span of time. You saw that the location of leaves on trees could affect which giraffes survived. Over many generations, longer-necked giraffes would be more likely to reach the uppermost leaves on tall trees. This might make them more likely to survive, reproduce, and pass their traits on to their offspring. If this were to happen, longer necks would be called an adaptation to the tall-tree environment.

Adaptations that make a species more successful are not always traits that make the species stronger, bigger, or faster. For example, some adaptations decrease the chances that a species will be eaten by another species. Adaptations of this type include the skin colors of lizards, the spines of porcupines, and the scent glands of skunks.

CHALLENGE

How do factors such as the environment and the presence of predators affect the process of natural selection?

THE TOOTHPICK WORM MODEL

Imagine that you are a bird that eats small worms. In this activity, toothpicks of two different colors will represent the worms that you eat.

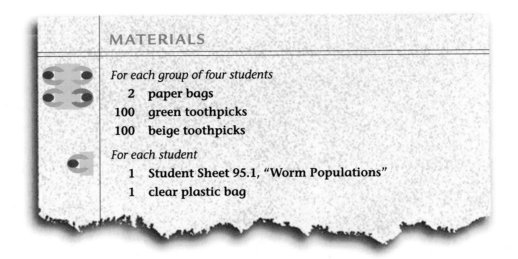

MATERIALS

For each group of four students
2   paper bags
100   green toothpicks
100   beige toothpicks

For each student
1   Student Sheet 95.1, "Worm Populations"
1   clear plastic bag

PROCEDURE

1. Label one of the paper bags "Worms" and the other "Reserve Toothpicks."

2. Each toothpick represents a worm. Count 25 green "worms" and 25 beige "worms" and place them into the paper bag labeled "Worms." This is the initial number of worms. These amounts are already marked for you in both tables on Student Sheet 95.1, "Worm Populations" (Table 1 is for green worms and Table 2 is for beige worms).

3. Place the rest of the toothpicks into the bag labeled "Reserve Toothpicks."

4. Shake the "Worms" bag to mix the worms.

5. As directed by your teacher, scatter the worms on the "ground."

6. You are going to play the role of a bird that eats worms. Each person in your group "eats" (picks up) the first 10 worms he or she sees and places them into the clear plastic bag, which represents the bird's stomach.

7. Count the total numbers of green and beige worms eaten by your group. Record these totals in Row 2 of each table on Student Sheet 95.1. Be sure to stay in the column for this generation.

8. *Some worms are still alive.* Subtract the number of worms that your group "ate" from the initial population in that generation. For example, if your group collected 18 green worms, there must be 7 green worms still alive on the ground (25 – 18 = 7). Record the numbers of surviving green and beige worms in Row 3 of each table on Student Sheet 95.1.

9. *Each living worm is reproducing.* On Student Sheet 95.1, multiply the numbers of green and beige worms still alive by 4. For example, if you had 7 green worms still alive, there would be a total of 28 green offspring worms (7 x 4 = 28). Record this number in Row 4.

10. Add one toothpick for each new green and beige worm into your paper bag labeled "Worms." For example, if your group had 7 green worms surviving on the ground, you would add 28 green toothpicks to the paper bag.

11. On Student Sheet 95.1, add Rows 3 and 4 of each table to calculate the final populations of green and beige worms. Record these numbers in Row 5 of each table. Record these same numbers in Row 1 in the columns for the *next* generation.

12. Repeat Steps 4–11 for Generation 2. If you have time, perform the simulation for further generations.

EXTENSION

Repeat the activity wearing a pair of sunglasses with green lenses. How are your results different?

ANALYSIS

1. **a.** Determine the ratio of green to beige worms in each generation. For example, the ratio of green to beige worms in Generation 1 is 25:25, or 1:1.

   b. Describe how the ratio of green to beige worms changed over the three generations.

   c. Why do you think this change occurred? Explain.

2. Imagine that you performed this simulation for another generation. What do you predict the ratio of green to beige worms would be? Explain your prediction.

3. Due to a drought, grass begins to dry out and die, leaving only dead grass stalks. What is likely to happen to the ratio of green to beige worms? Explain.

4. **a.** In this activity, what effect did the environment have on the process of natural selection?

   b. In this activity, what role did the predator (bird) have in the process of natural selection?

5. **Reflection:** Why do you think earthworms are beige and not green?

MODELING

During the history of Earth, species have both evolved and become extinct. Why do some species survive to reproduce while others do not?

CHALLENGE

What role does variation play in the process of natural selection?

Why do these four different species of birds have such different beaks?

THE FORKBIRD MODEL

In this activity, you will role-play a single species called "forkbirds." Forkbirds feed by either spearing or scooping their food. During feeding time, each bird gathers "wild loops" and immediately deposits them in its "stomach" before gathering more food. Your goal is to gather enough food to survive and reproduce. This will allow you to pass your genes on to another generation. Occasionally, a forkbird offspring will have a genetic mutation that makes it look different from its parent.

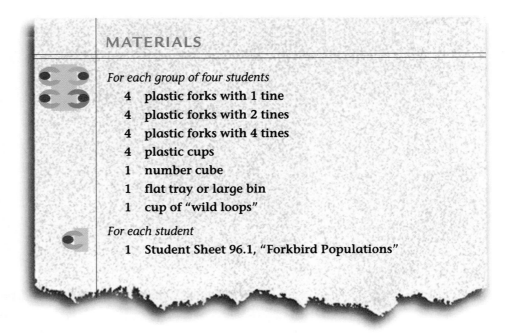

MATERIALS

For each group of four students

- 4   plastic forks with 1 tine
- 4   plastic forks with 2 tines
- 4   plastic forks with 4 tines
- 4   plastic cups
- 1   number cube
- 1   flat tray or large bin
- 1   cup of "wild loops"

For each student

- 1   Student Sheet 96.1, "Forkbird Populations"

PROCEDURE

1. The initial forkbird population has beaks with only two tines. Each person in your group should begin the activity with a 2-tined fork. Record the initial population of each type of forkbird in Table 1 of Student Sheet 96.1, "Forkbird Populations."

2. Your teacher will tell you when feeding time begins, and then all of the forkbirds can feed.

3. When feeding time ends, count the number of wild loops eaten by each forkbird. Within your group, the two forkbirds that gathered the most food survive to reproduce. (If there is a tie for second place, then three forkbirds survive. The two forkbirds that tie should keep their forks and skip Step 4.)

4. The two surviving forkbirds should each toss the number cube. Use the table below to determine the type of beak of the offspring of each surviving forkbird. The group members whose forkbirds did not survive should now assume the roles of the offspring.

Number Cube Key	
Your Toss	**Forkbird Offspring**
1	1-tined forkbird
2	2-tined forkbird
4	4-tined forkbird
3, 5, 6	same as parent forkbird

5. Record the new population of each type of forkbird in your group in the next row on Student Sheet 96.1.

6. Return all of the wild loops to the "forest floor" (tray or bin) to simulate the growth of wild loops.

7. Repeat Steps 2–6 for nine more rounds to represent additional generations.

8. Share your data with the class. As a class, record the population of each type of forkbird over many generations. Be sure to copy the class data onto Student Sheet 96.1.

9. Create a graph of the class totals of each type of forkbird over many generations. You can plot the data for all three types of forkbirds on a single graph. Be sure to title your graph, label your axes, and provide a key.

ANALYSIS

1. Which type of forkbird was the most successful? Explain how the class data support this conclusion.

2. **a.** Look at your graph of the class results. Describe what happened to the number of each type of forkbird over many generations.

   b. In the forkbird model, mutations at reproduction were much more common than they are in real life. Imagine that the number of mutations was lowered, so that the vast majority of offspring had beaks similar to those of their parents. Predict what you think would have happened to the numbers of each type of forkbird in future generations.

3. How did the forkbird activity simulate the process of natural selection? Explain.

4. The forkbirds that you studied are a single species. Although they look slightly different, they are part of a single, interbreeding population. Imagine that a change in the food supply occurred.

   a. As a result of heavy rains, the major source of forkbird food is now soft berries, like blueberries. After many, many generations, how many types of forkbirds do you think will be in the population? Explain your reasoning.

   b. As a result of a drought, the major source of forkbird food is now sunflower seeds. After many, many generations, how many types of forkbirds do you think will be in the population? Explain your reasoning.

5. **Reflection:** The cheetah, an extremely fast and efficient hunter, is an endangered species. The few cheetahs alive today show very little variation. How does this help to explain why cheetahs are on the verge of becoming extinct?

READING

In Activity 92, "Time For Change," you saw that the types of living organisms have changed throughout Earth's history. Where do all of the new types of organisms come from?

CHALLENGE

What role do mutations play in natural selection?

Three different species of bears

READING

In the Ecology unit, you learned that each species has a particular role within its ecosystem. The angelfish is adapted to eating small aquatic worms. To people, most adult angelfish of a particular breed appear the same: they are all of similar size and coloration and eat the same types of food. But there is some variation—every angelfish is slightly different (Figure 1). Consider other organisms that you might think are identical. What could you do to identify differences among individuals within the species?

One way to look for variation is to examine physical features, such as color and shape. Often, features like the width or pattern of stripes on an angelfish are slightly different from one fish to the next. Since some physical differences are due to genetic differences, they can be passed along through the generations.

Figure 1: Variation Between Two Angelfish

STOPPING TO THINK 1

Think about similarities and differences among ten different people you know.

a. What are some physical features that are likely to be a result of genetic differences?

b. What are some physical features that may not be a result of genetics, but a result of some other factor(s), such as development from birth to adulthood?

c. What are some physical features that might be a result of both genetics and other factors?

In Activity 96, "Battling Beaks," you modeled a forkbird population that showed variation. Although all the forkbirds were from the same species, there were 1-tined, 2-tined, and 4-tined forkbirds. What was the source of these differences?

All genetic variation exists because of **mutations**. The reproduction of the genetic material does not always happen perfectly. As a result, occasionally an offspring has features that do not exist in the parents or even in the rest of the species. Some mutations are harmful. For example, a bird might be born with a beak of such unusual shape that the bird cannot feed. Such mutations are not passed on to the next generation, since the affected organism does not survive to reproduce.

In many cases, a mutation is neither helpful nor harmful. The 1-tined forkbird from the previous activity was an example of this type of mutation. Even though it was not as successful as the 4-tined mutation, the 1-tined beak was neither helpful nor harmful when compared to the 2-tined beak. Since there was no advantage or disadvantage to this type of beak, the 1-tined forkbird did not die out in the population.

STOPPING TO THINK 2

Imagine that you own a dog that recently gave birth to a litter of puppies. Your veterinarian informs you that one of the puppies has a genetic mutation.

a. Think of a mutation that the puppy could have that would be neither helpful nor harmful.

b. Think of a mutation that the puppy could have that would be harmful.

In some cases, a mutation is helpful. Imagine that a bird from a species that eats small nuts is born with a larger beak than the rest of the population. The larger beak allows this bird to eat large nuts as well as smaller nuts. If nuts became harder to find, this mutation could help this bird survive and reproduce. Any larger-beaked off-spring might continue to be more successful than the rest of the bird population. After many generations, all of these birds might have larger beaks (Figure 2). In the previous activity, the 4-tined forkbird was an example of a helpful mutation.

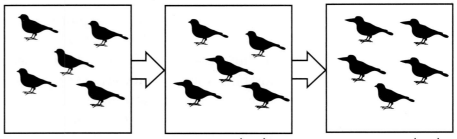

many generations later   many more generations later

Figure 2: Evolution of Larger-Beaked Birds
As a result of a helpful mutation occurring just once, an entire population of birds might look very different after many generations.

In a new environment, natural selection might favor a mutation that is not favorable in the original environment. If this population eventually can no longer successfully reproduce with the population it came from, it is considered a different species.

STOPPING TO THINK 3

You may have heard someone who is wrapping a present say, "I wish I had another hand!" Explain why an organism cannot choose to have a mutation that would enable it to live more successfully in its environment. For example, could birds choose to have larger beaks? Explain your reasoning.

Figure 3: The Galapagos Islands

The fossil record provides evidence that many different species have lived during the history of Earth. But Charles Darwin was one of the first people to notice that living species also provide evidence for evolution. In the late 1830s, Darwin traveled on a ship called the *Beagle* that sailed around the world. He collected evidence and made careful observations of the natural world wherever the voyage took him. One of the places that the ship stopped was the Galapagos Islands, a chain of islands located in the Pacific Ocean, west of South America (Figure 3).

In the Galapagos Islands, Darwin collected samples of many different species, including 14 species of finch (a small bird). All of the finches were similar, but the species varied in color, size, and beak shape. Darwin observed a relationship between the shape of a finch's beak and the food that it ate. Scientists had noticed that the beak of each species was particularly well-adapted to getting a specific type of food, such as a certain seed or insect (Figure 4).

Cactus finch   *Large ground finch*   *Warbler finch*

Figure 4: A Few Galapagos Finches

Based on his observations, Darwin hypothesized that all 14 different finch species had evolved from one single ancestral species. He proposed that, thousands or even millions of years ago, a single species of South American finch migrated and began nesting on the islands. Over many generations, different adaptations proved more successful on one island than on another. Because each island is separated by some distance from others in the chain, the finch population on each island is relatively small and isolated. This allowed helpful genetic mutations to spread within a population—by natural selection—more quickly than usual. Eventually, changes in beak shapes, combined with the spread of other helpful mutations, resulted in enough differences that the various finches became separate species, each adapted for a different ecosystem role.

Today, scientists use genetic evidence to compare similarities and differences among species. By testing the genes of the various finches, scientists have shown that the finches are very closely related, providing more evidence that Darwin's hypothesis is correct.

STOPPING TO THINK 4

Darwin identified 14 species of finch on the Galapagos Islands. Your friend says that this means only 14 mutations occurred within the finch populations. Explain whether you agree with your friend and why.

But you don't need isolated islands to produce new species. Remember the Nile perch of Lake Victoria in Africa, which you studied in the previous unit? One consequence of the introduction of these large fish into the lake was the extinction of up to 200 species of just one type of fish—the cichlid.

Different species of cichlids

How did so many species of the same fish family ever come to exist in a single lake? A single lake provides a surprising number of different places to live and ways to survive. Differences in the amount of light, wind, mud, sand, temperature, plants, predators, and insects produce a variety of habitats within one lake. Lake Victoria provides so many different habitats that over 300 different species of cichlids had evolved within the lake before the introduction of the Nile perch.

Are all of these cichlids really descended from a single ancestor? Every line of evidence suggests this is so. Modern genetic evidence indicates that all the cichlids in Lake Victoria evolved from a common ancestor within the last 200,000 years. That's a short period of time in terms of evolution!

ANALYSIS

1. Are mutations always helpful? Explain.

2. How can mutations enable the evolution of a new species to occur? Use the story of the cichlids to help you explain your ideas.

3. Under ideal conditions, bacteria have a generation time of about 20 minutes. Humans have a generation time of about 20 years. Which would you expect to evolve faster? Why?

INVESTIGATION

Fossils have been found in Precambrian rocks 3.5 billion years old. But most have been found in rocks of the Paleozoic, Mesozoic, and Cenozoic eras, which are all less than 550 million years old. The types of organisms found in different rocks can provide important information about the history of life on Earth. The term **fossil record** refers to all of the fossils that have been found on Earth.

The fossil record has been used to classify fossils into families. A family is a category smaller than a kingdom, phylum, class, or order, but larger than a genus or species. For example, dogs are in the family Canidae, which also contains foxes, jackals, coyotes, and wolves. Lions are in the same kingdom, phylum, class, and order as dogs, but they are in a different family: Felidae. This family includes leopards, tigers, cheetahs, house cats, and extinct species such as the saber-toothed cat. You will investigate how the numbers of families in the fish, mammal, and reptile classes have changed over geological time.

CHALLENGE

What can you learn about evolution by comparing the fossil records of fish, mammals, and reptiles?

Classifying Carnivores		
Classification Level	**Dogs**	**Lions**
Kingdom	Animalia	Animalia
Phylum	Chordata	Chordata
Class	Mammalia	Mammalia
Order	Carnivora	Carnivora
Family	**Canidae**	**Felidae**
Genus	*Canis*	*Panthera*
Species	*familiaris*	*leo*

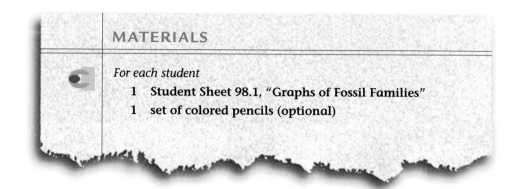

MATERIALS

For each student
1   Student Sheet 98.1, "Graphs of Fossil Families"
1   set of colored pencils (optional)

PROCEDURE

1. Table 1 below provides the history of all the families of fish currently known from the fossil record. When a fossil is found that does not belong to any family found in *earlier* geologic time periods, we call it a "first appearance." It is the first appearance of that family in the fossil record. When a fossil is found that does not belong to any family found in *later* geologic time periods, we call it a "last appearance." It is the last appearance of that family in the fossil record. Look at Table 1 and discuss the following questions with your partner:

- Between which years did the greatest number of fish families appear in the fossil record? In what era was this period of time?

- Between which years did the greatest number of fish families disappear from the fossil record? In what era was this period of time?

Table 1: History of Fossil Fish Families									
Era	**Precambrian**	**Early Paleozoic**		**Late Paleozoic**		**Mesozoic**			**Cenozoic**
Time (mya)	>545	485	425 365	305	245	185	125	65	0
Number of first appearances	0	25	43 162	67	13	52	33	84	299
Number of last appearances	0	9	31 158	49	48	36	20	44	34

➤

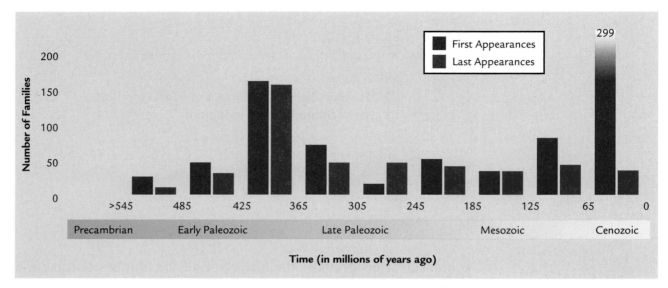

Figure 1: Graph of Fossil Fish Families Over Time

2. Figure 1 presents a double bar graph of the data shown in Table 1. Look at the graph and discuss with your partner in what ways the graph makes the data easier to interpret.

3. Use the information in Table 2 to make a double bar graph for families of reptiles, similar to the one for fish shown in Figure 1. Since you will be comparing graphs, be sure to use the same scale on the y-axis.

Table 2: History of Fossil Reptile Families									
Era	**Precambrian**	**Early Paleozoic**			**Late Paleozoic**		**Mesozoic**			**Cenozoic**
Time (mya)	>545	485	425	365	305	245	185	125	65	0
Number of first appearances	0	0	0	0	3	67	95	68	97	35
Number of last appearances	0	0	0	0	1	57	93	46	84	26

Table 3: History of Fossil Mammal Families

Era	Precambrian	Early Paleozoic		Late Paleozoic			Mesozoic			Cenozoic
Time (mya)	>545	485	425	365	305	245	185	125	65	0
Number of first appearances	0	0	0	0	0	0	6	14	33	404
Number of last appearances	0	0	0	0	0	0	2	8	33	262

4. Use the information in Table 3 to make a double bar graph for families of mammals, similar to the one for fish shown in Figure 1. Since you will be comparing graphs, be sure to use the same scale on the y-axis.

A familiar example of a fossilized reptile

ANALYSIS

1. **a.** Use the graphs to place the three different classes in order, based on when they first appeared in the fossil record.

   b. What could this order tell you about the evolution of these types of species?

2. **a.** What are some possible explanations for the disappearance of a family from the fossil record?

   b. How could Darwin's theory of natural selection explain the disappearance of these families?

3. What could explain the appearance of a family in the fossil record?

4. **a.** The Cenozoic Era is often referred to as the "Age of Mammals." Using evidence from this activity, explain why.

   b. Based on evidence from this activity, what could you call the Mesozoic Era? Explain your reasoning.

   c. Look at the appearances and disappearances of families over time on all three graphs. Why is it misleading to label an era as the "age of" any particular class?

INVESTIGATION

Whales, dolphins, and porpoises are mammals that live in the sea. Like all mammals, they are warm-blooded animals that give birth to live young and need air to breathe. DNA evidence shows that whales are closely related to hoofed land mammals such as hippopotamuses, pigs, cows, and sheep. All of these mammals are thought to have descended from a single species that lived millions of years ago and is now extinct. Besides DNA evidence, what other evidence suggests that these animals are related?

CHALLENGE

How are modern and fossil skeletons used to investigate evolution?

THE FOSSIL EXHIBIT

You've just been hired as the assistant curator of the fossil collection of a museum. On your first day, you discover that the skeletons in the exhibit on the evolution of whales have all been moved to a new room and need to be arranged. Unfortunately, you are not a whale expert and the skeletons are not clearly labeled.

A local middle school has scheduled a field trip to the museum. It is very important that you arrange the skeletons properly before the students arrive. You decide to examine them to see if you can figure out how they should be arranged.

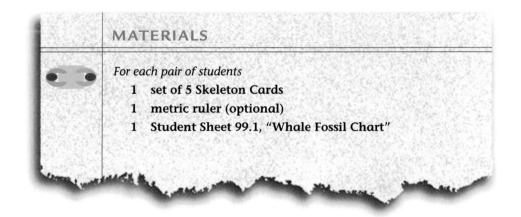

MATERIALS

For each pair of students

1   set of 5 Skeleton Cards
1   metric ruler (optional)
1   Student Sheet 99.1, "Whale Fossil Chart"

PROCEDURE

1. Compare the five Skeleton Cards. Based on similarities you observe, group the skeletons into two sets, each containing two or three cards. The set of skeletons containing Skeleton A should be called "Group 1." The other set of skeletons will be "Group 2."

Comparing Skeletons

	Similarities	Differences
Group 1 skeletons: A, _____		
Group 2 skeletons: _____		
Group 1 skeletons compared with Group 2 skeletons		

2. Create a table in your science notebook like the one shown above. In the first column, record which skeletons you put in each group.

3. Compare the skeletons *within* each group. In your table, describe and record as many similarities and differences as you can.

4. Compare Group 1 skeletons with those of Group 2. In your table, describe and record as many similarities and differences as you can.

5. *It's time to figure out how to arrange the exhibit!* Use similarities and differences in the skeletons to arrange the cards in order. (While all five skeletons can be in a single line, they don't have to be.) Record the order in which you have arranged the skeletons. **Hint:** Place the two least similar skeletons on either side of your desk. Then arrange the other three skeletons between them.

6. *You're in luck! You discover a chart with information about the relative ages of the five skeletons.* Collect Student Sheet 99.1, "Whale Fossil Chart," from your teacher.

7. Compare the age data from Student Sheet 99.1 with the order in which you placed the skeletons in Step 5. If necessary, rearrange your Skeleton Cards. Record your final reconstruction of the museum exhibit in your science notebook.

➢

ANALYSIS

1. **a.** What kinds of skeletal changes appear to have occurred during the evolution of whales?

   b. What can you infer about the changes in habitat that occurred at the same time as these skeletal changes?

2. Use natural selection to explain how these changes (or one of these changes) could have occurred.

3. In this activity, you examined extinct and modern whale skeletons. How does the study of these skeletons provide evidence about how species are related?

4. Look again at Skeleton A. This is known as an ambulocetid (am-byoo-low-SEE-tid). The word *ambulocetid* means "walking whale." Where do you think the ambulocetids lived? Describe how you think they lived.

EXTENSION

Find out more about current research on whale evolution. Start at the SALI page of the SEPUP website.

Scientists use evidence such as similarities in skeletal structures and other physical traits to investigate evolutionary relationships. Thanks to advances in genetics and biotechnology, scientists studying evolution can now also use the genetic material itself.

Each cell in an organism contains the genetic information needed to perform all its functions, such as obtaining energy, moving, and getting rid of wastes. You may already know that the genetic information is located in DNA in the chromosomes found in every cell. DNA is made up of four chemicals whose names are abbreviated as A, T, G, and C. These chemicals are strung together like beads on a string. Differences in their order result in different messages. Think of these chemicals as letters in an alphabet: there are only 26 letters in the English alphabet, but there are millions of words. In a similar way, DNA contains millions of biological messages. The more similar the DNA, the more similar the messages used to run the organism.

CHALLENGE

How does DNA provide evidence about how animals are related?

These scientists are analyzing evidence from DNA.

THE COMMON THREAD

You are an evolutionary biologist investigating relationships among different species. When a geneticist you work with offers to provide you with DNA samples from various animals, you do some background research. You find out that the samples are from a gene that is similar in all vertebrates. This means you can easily use them to compare species of vertebrates.

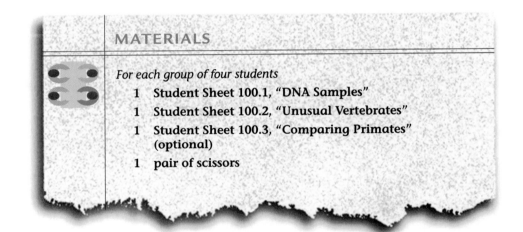

MATERIALS

For each group of four students

1   Student Sheet 100.1, "DNA Samples"
1   Student Sheet 100.2, "Unusual Vertebrates"
1   Student Sheet 100.3, "Comparing Primates" (optional)
1   pair of scissors

PROCEDURE

Part One: Comparing Vertebrate Classes

1. Compare the five samples of DNA on Student Sheet 100.1, "DNA Samples." With your partner, discuss any similarities or differences that you notice among the samples.

2. Use the DNA samples to determine whether animals in the same class have more similarities in their DNA with one another than they do with animals in other classes. Record your ideas in your science notebook. **Hint:** First compare just two animals and count the number of differences in their DNA. You can make a table like the one shown on the next page to record these counts.

Numbers of Differences Between DNA Sequences

	Mammal #1	Mammal #2	Mammal #3	Reptile
Fish				
Mammal #1	—-			
Mammal #2	—-	—-		
Mammal #3	—-	—-	—-	

3. In Activity 99, "A Whale of a Tale," you examined evidence that whales are mammals. Look again at your DNA samples. Discuss with your group whether these samples provide additional evidence that whales are mammals.

Part Two: Gathering More Evidence

4. In Activity 76, "People, Birds, and Bats," you classified a number of vertebrates into different classes. Review how you classified four of these animals: the kiwi, platypus, armadillo, and bat.

5. A local biotechnology center provides you and the geneticist with DNA samples from these four animals. Use Student Sheet 100.2, "Unusual Vertebrates," to compare the DNA samples of these four animals with the DNA samples from Part One. In your science notebook, create a table similar to the one above to record your comparisons.

6. In your science notebook, record whether the DNA evidence supports or conflicts with the way that you had classified these animals. If you make any changes to your classification, be sure to record them.

ANALYSIS

1. In this activity, you used DNA to evaluate relationships among animals. How does DNA provide evidence about how species are related?

2. Would you expect the DNA of a seahorse to be more like the DNA of a horse or the DNA of a trout? Use evidence from this activity to support your answer.

3. **a.** Look back at the evolutionary tree in Figure 2 of Activity 89, "Here Today, Gone Tomorrow?" Draw a simple tree that shows the evolution of reptiles, fish, and mammals.

   b. Explain how DNA evidence helps you draw evolutionary trees.

4. The first mammals evolved from a reptilian ancestor, 200 million years ago. Explain why it is not accurate to say that humans evolved from lizards.

EXTENSION

Compare the human, chimpanzee, and rhesus monkey DNA sequences provided on Student Sheet 100.3, "Comparing Primates." Use this evidence to draw an evolutionary tree for these three types of primates.

By comparing fossil evidence with living species, it is clear that almost all the species that have ever lived on Earth have become extinct. As this diagram shows, most living species are descended from a small fraction of the species that have ever existed.

Wherever a "branch" of this evolutionary tree of species ends, an extinction occurred (except at the present day).

Why do some species survive while others disappear? Species die out for many reasons. These include environmental change, competing species, habitat loss, and disease. Human activity can contribute to each of these causes.

CHALLENGE

How does natural selection help explain the extinction of the dodo bird and the success of the common pigeon?

➤

PROCEDURE

Compare the fate of the dodo bird with that of the common pigeon.

RELATED BIRDS, DIFFERENT FATES

The common pigeon seems to be everywhere—almost everyone has seen one of these birds. No one alive today has seen a dodo bird, and no preserved specimens of this extinct species exist. There are 27 orders of birds. Based on skeletal comparisons, the dodo and the pigeon are classified in the same order. The pigeon and the dodo are evolutionary cousins!

The Dodo Bird

Often portrayed as flightless, fat, slow, and stupid, the dodo bird (*Raphus cucullatus*) has become a symbol for something out-of-date or clumsy. Some people think it somehow fitting that the dodo species went extinct. How could natural selection have produced such a creature in the first place?

Dodos lived successfully for several million years on the island of Mauritius in the Indian Ocean (Figure 1). Migratory birds probably had settled on Mauritius long before, just as Darwin's finches did on the Galapagos Islands. Contrary to popular belief, evidence shows that the flightless dodo was a slender, fast-running animal (Figure 2). Although it competed for resources with many other bird species, the 30- to 50-pound dodo had few predators on the island. Without predators, dodos could nest on the forest floor and eat fruit that fell from trees. Flight was unnecessary for survival and so, over many generations, the new species evolved to become flightless.

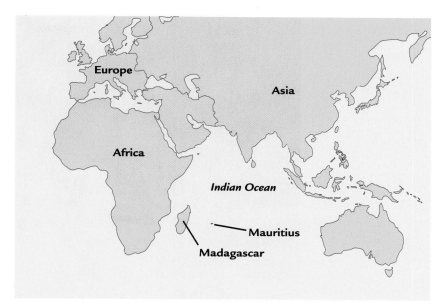

Figure 1: Location of Mauritius

Mauritius is a volcanic island about 10 million years old, about 500 miles east of Madagascar. Today, it is an independent country with a population of over 1 million people.

Figure 2: What Remains of the Dodo

Made of bones gathered from the island during the 1850s, this skeleton confirms that the dodo was flightless, but not that it was slow-moving.

Penguins, as well as the kiwi (shown on page E-25), are examples of living species of flightless birds.

In 1505, Portuguese sailors became the first mammals to set foot on Mauritius. Soon, the island became a common stopping place for ships travelling between Europe and Asia. Because of its large size and inability to fly, the dodo became a hunting target for hungry sailors. Because dodo nests were on the ground, their eggs were easily found and eaten by the rats, pigs, monkeys, and other animals that accompanied the sailors. In addition, human settlers' need for cleared land and wood greatly reduced the size of the dodo's forest habitat. In 1681, less than 200 years after the first predators arrived on Mauritius, the last dodo bird was killed.

The Common Pigeon

Native to Europe and Asia, pigeons now thrive on five continents. The common pigeon, or rock dove (*Columba livia*), was first domesticated by humans between five and ten thousand years ago. Early humans raised the birds for food, and pigeon meat is still a delicacy in many cultures. Later, pigeons were bred to race, to deliver messages, to do stunts, and for show (Figure 3).

Perhaps even before becoming domesticated, pigeons discovered that human structures were convenient, safe places to nest. In addition, fields and marketplaces provided an easy-to-gather, year-round food supply. During their several thousand years of close association with humans, human-bred pigeons have escaped and mated with wild pigeons, sharing genes with them. As a result, pigeon populations found near people, known as feral pigeons, are quite different from wild pigeons. They can fly faster and for longer distances, breed earlier in life, produce more offspring, and live at a much higher population density. The remaining population of wild pigeons is decreasing, and may soon dwindle to zero. Meanwhile, the population of feral pigeons continues to grow.

Figure 3: Pigeon Diversity
Over many generations, through both natural processes and breeding, the pigeon species has evolved adaptations to many successful lifestyles associated with the human species.

ANALYSIS

1. If humans had never interacted with either the dodo or the pigeon, how do you think the history of each species would be different? Explain your reasoning.

2. Could the evolution of feral pigeons be described as the formation of a new species? Explain.

3. Use natural selection to explain how the flying bird that first settled on Mauritius might have evolved into the flightless dodo. In your answer, be sure to include the role of mutations.

4. Your friend argues that the dodo bird became extinct because it was a poorly adapted species, destined for failure. Do you agree? Explain.

5. Imagine that advances in science and technology allow genetic engineers to re-create living dodo birds and mammoths.

   a. Should mammoths be re-created and released into the Arctic ecosystem? Support your answer with evidence and discuss the trade-offs of your decision.

   b. Should dodos be re-created and released into the ecosystem of modern Mauritius? Support your answer with evidence and discuss the trade-offs of your decision.

   Hint: To write a complete answer, first state your opinion. Provide two or more pieces of evidence that support your opinion. Then discuss the trade-offs of your decision.

EXTENSION

Find out more about extinct and endangered species. Start at the SALI page of the SEPUP website.

Using Tools and Ideas

Unit G

Using Tools and Ideas

Meera and Justin could always be found using the computer. Today they were online, looking up opinions and information about the new surgical techniques they had learned about in school that day. Their grandmother came over to watch.

"Grandma, is it true that when you started out as a computer programmer, a computer covered the entire top of a desk?" asked Meera.

"Actually," their grandmother replied. "The first computer I ever used filled an entire room!"

Justin was impressed, but also puzzled. "No way! Why were computers so big? They are so much smaller now. Why weren't they able to do that back then?"

"That's a good question, Justin," said his grandmother. "New technologies, like microprocessors, had to be invented first. In fact, it wasn't until the 1970s that the first personal computers were invented."

"Microprocessors? What are those?" asked Meera as she began to search online.

"I don't know, but I'd sure like to find out how they figured out how to make computers!" said Justin.

• • •

How are new technologies such as computers invented? How is the process of invention related to science? How are new solutions to problems in life science developed? In this unit, you will develop and design your own tools and strategies related to human health and safety. What problems would you like to solve with your new skills?

INVESTIGATION

Have you ever thought of yourself as an inventor? You've probably invented solutions to many problems that you face. But part of being an inventor involves recognizing and describing your solutions. How would you handle the problem of a broken arm? What solutions could you describe?

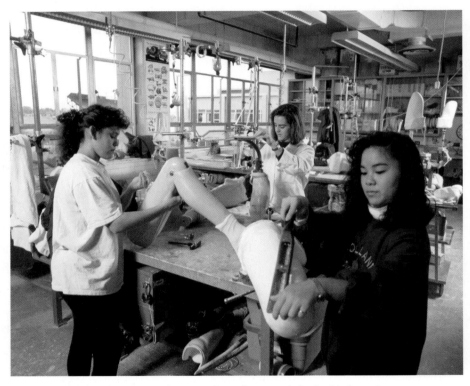

These university students are making prosthetic devices to help disabled people.

THE BROKEN ARM

Yesterday you fell while in-line skating down a hill. Your arm is broken pretty badly and is in a cast from the middle of your upper arm to the first joint of your fingers. The pain and swelling make your fingers useless, which the doctor says will continue for a couple of weeks. And to make matters worse, it's the hand you normally write with! What problems will you face in carrying out the activities of daily life? How can a scientific approach help you to design useful solutions?

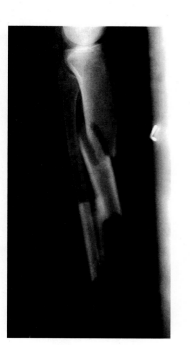

In this activity, you will simulate having a broken arm. As you try to perform various tasks, you must assume that you are alone and cannot ask for help. Since your arm and hand are swollen and in pain, you cannot use any part of the arm or hand that is in the sling.

CHALLENGE

What tools and strategies can you invent to deal with a broken arm?

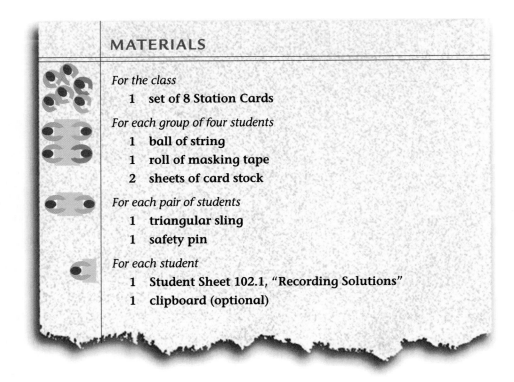

MATERIALS

For the class

   1   set of 8 Station Cards

For each group of four students

   1   ball of string

   1   roll of masking tape

   2   sheets of card stock

For each pair of students

   1   triangular sling

   1   safety pin

For each student

   1   Student Sheet 102.1, "Recording Solutions"

   1   clipboard (optional)

 **SAFETY**

Some of the activities involve eating. Be careful not to contaminate any of the food. Be sure to follow your teacher's instructions about what to do with any used utensils or food containers.

PROCEDURE

1. With your partner, decide who will wear the sling first. One of you will wear the sling to the first half of the stations; the other partner will wear the sling to the other half of the stations.

2. Have the first person simulate a broken arm by putting the dominant arm (the one usually used for writing) in a sling. Look at Figure 1 to find out how to make a sling.

a.

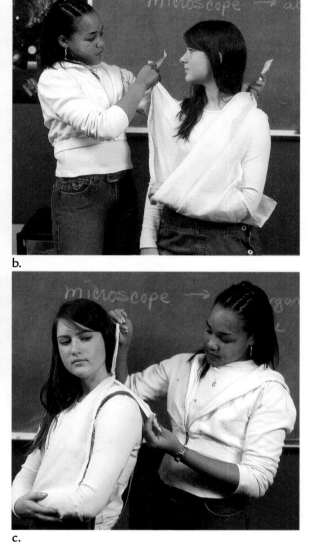

b.

c.

Figure 1: Tying A Sling

a. Lay triangular bandage under arm with the point toward the elbow.

b. Bring up the bottom corner to the other side of the neck.

c. Tie behind the neck.

3. Go to each station as directed by your teacher and try to perform the task described on the Station Card. You will have to devise a solution to each problem. You can either try to develop a strategy or use the supplies provided to make a simple tool. Try to invent a solution that will allow you to carry out your daily activities independently.

4. As you complete each task, have your partner use Student Sheet 102.1, "Recording Solutions," to record the tools and strategies you use.

5. After you have visited half of the stations, have your partner wear the sling and repeat Steps 3 and 4.

➢

ANALYSIS

1. **a.** Describe the most challenging problem you think you would face in everyday life if you broke your arm.

   b. Use both a written description and a diagram to explain how you would solve this problem.

2. What were the strengths and weakness of putting your arm in a sling to model what it would really be like to break your arm?

   Hint: You may want to interview some people who have dealt with real injuries before answering this question. Consider pain and attitudes of others in your answer.

 3. Even without broken arms, people in all professions, including cooks, cleaners, surgeons, occupational therapists, and plumbers, need to invent solutions to problems in the course of their jobs. Choose a profession and give an example of a kind of problem someone in that profession may need to solve.

 4. **a.** Identify a problem you would like to solve.

   b. Describe a tool you would like to invent to solve this problem.

   c. How would you start to make the tool you described?

READING

You have investigated how you would use tools and strategies to solve problems you would have if you couldn't use of one of your arms. People invent tools or strategies to solve many kinds of problems. For example, glasses were invented to help people who have vision problems to see better. Telescopes were invented to help people see distant objects. Microscopes help people see very tiny objects. Read the cases presented here to learn about tools and strategies used by people who do not have the use of all their body parts.

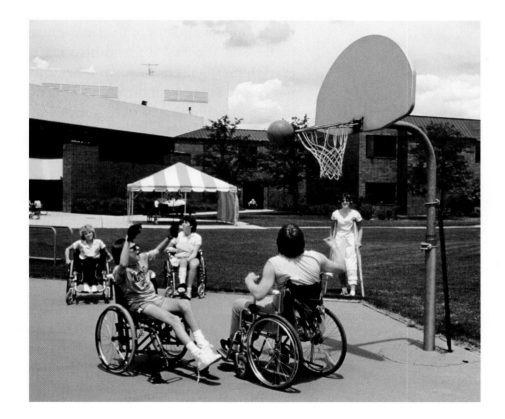

CHALLENGE

How do people with disabilities use tools and strategies to accomplish their daily activities?

PROCEDURE

Read the case studies below. Think about what tools and strategies each person uses to solve problems in real life.

Case 1

Sarah is a high school student who injured her back in a serious car accident last month. Her doctors predict she will be fine in a few months if she sticks to her physical therapy and takes good care of her back. One of the hardest parts of Sarah's day is the early morning: her back hurts when she tries to bend over, which makes showering and getting dressed a painful process. Sarah uses a piece of string to hang her shampoo bottle from the showerhead so she can easily reach the bottle. She also had a friend help her organize her clothing so that everything she needs is either in her top dresser drawer or hanging in her closet. But she's most proud of her "shoe stick": she used strong tape to attach a shoehorn to a meter stick. This allows her to slip on her shoes without bending over.

STOPPING TO THINK 1

What tools or strategies does Sarah use to help her with her morning routine?

Case 2

Aimee Mullins was born without the fibula in both of her legs. The fibula is one of the two long bones in the lower leg. When Aimee was a baby, her parents made the difficult choice to have her legs amputated just below the knees. This would give her the possibility of learning to walk with artificial legs. By the age of two, she had learned to walk with heavy wooden legs, called **prostheses** (prahs-THEE-sees). Aimee now uses several different pairs of prosthetic legs. She has different pairs for sports such as swimming and running, a pair for everyday use, as well as a cosmetic pair that look and feel like real legs. When she was a young child and teen she danced, played soccer, skied, and biked. Now she is a Paralympic runner (her record in the 100-meter dash is 15.77 seconds) and a fashion model. Aimee appreciates her parents' attitude toward her disability: "They treated me like they would any other kid. They made me believe that I could do whatever I set my mind to."

STOPPING TO THINK 2

a. What tools did Aimee use to be able to walk, swim, and run?

b. What strategies did Aimee's parents use to help Aimee achieve her goals?

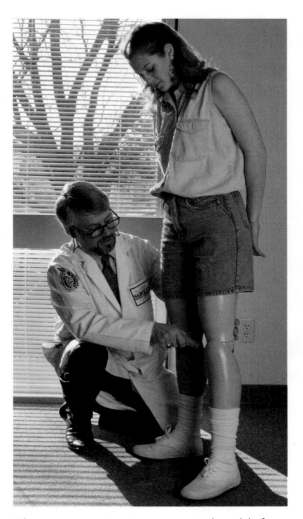

This young woman is trying on a research model of a prosthetic leg.

Case 3

Without the new technologies of the past thirty years, Aimee would not be able to have such useful artificial legs. Though the first artificial leg was constructed around 300 B.C.E., the only ones available until very recently were "peg legs," large wooden sticks that supported the body. Eventually engineers developed new materials and began to build legs with hinged knees. One new approach is to carefully drill a titanium implant into the bone of the remaining part of the leg, which then grows around the implant, attaching the prosthesis permanently. Others are working on a prosthetic leg that connects to the wearer's own nerves.

Hugh Herr uses his climbing legs.

Hugh Herr, a researcher at the Massachusetts Institute of Technology, lost his lower legs due to frostbite when he was mountain climbing at age 17. He engineered a leg that allows him to continue climbing (see photo below). Now he is working on an artificial leg that will provide thrust and also adjust its movement based on pressure and speed. This will help users, both athletes and non-athletes, to adjust to different surfaces and move more naturally.

STOPPING TO THINK 3

a. What problems are addressed by improvements in artificial leg technology?

b. What problems are not addressed by these advances?

Case 4

Artificial vision is more difficult to develop than limbs, since it involves sending complex signals from the eyes to the brain. Even so, scientists are working on it. In the meantime, the Americans with Disabilities Act requires that efforts be made to provide equal access to buildings and other facilities for blind people and people with limited vision. Audio and other technologies, for instance, help all people to use computers and other devices. However, some real-world tasks are not solved through modern technology, but rather through practice and concentration.

Brandon is a blind 14-year-old with two older brothers and a younger sister. All of Brandon's siblings and both his parents can see. When Brandon was 12, all the boys in his class from school were invited to go on a camping trip for a friend's birthday. Brandon very much wanted to go, but he had never been camping before. He heard that his friend's father was going to teach everyone how to safely build and light a campfire. But Brandon didn't even know how to light a match. He asked his mother if he could practice lighting a match so that he could help with the campfire on the trip. Brandon's mother wasn't sure that she wanted her son lighting matches on his own, but she agreed that he could practice using a match to light candles over the kitchen sink. She agreed to stay with him until he could safely accomplish this by himself.

With his fingers, Brandon examined a match. He noticed that one end was round. His mother explained that this was the end that would light. Brandon discovered that he could easily hear the sound of the flame when it lit and could slide his finger away from the flame. He practiced lighting candles by moving the lit match toward the candle. He could feel when he touched the wick with the match. He held the match still for a second while the wick caught fire. Then he moved the match away from the candle and blew out the match. After Brandon learned to light a match, his father let him learn how to light a fire in the fireplace. With his new skills, Brandon went on the camping trip and volunteered to light the campfire.

➢

STOPPING TO THINK 4

a. What tools did Brandon use to learn to light a campfire?

b. What strategies did Brandon and his parents use to help him achieve his goals?

ANALYSIS

1. a. Suggest an alternative way to solve one of the problems described in the cases.

b. Describe at least three other examples of ways in which technology helps people overcome physical limitations.

2. Reflection: Describe a problem that you face and how you might solve it using either a tool or a strategy.

PROJECT

New tools and procedures in the field of medicine are helping people live longer, healthier lives. From eyeglasses and artificial legs to modern surgical techniques, technology is helping more and more people to improve the quality of their lives.

Doctors and bioengineers have worked together to design and engineer artificial heart parts. For example, artificial aortas and valves can now replace damaged structures. The aorta is the main artery carrying blood from the heart. Blood is stopped from flowing back into the heart by the aortic valves. In patients with a genetic condition known as the Marfan syndrome, the aorta often swells. When this happens the aortic valves no longer meet and blood can flow backward into the heart. This is a serious condition, but it can be treated if the swelling is detected in time. If it is not treated, the aorta can swell so much that it bursts.

In this activity, you will design and test simple versions, known as **prototypes**, of a heart valve.

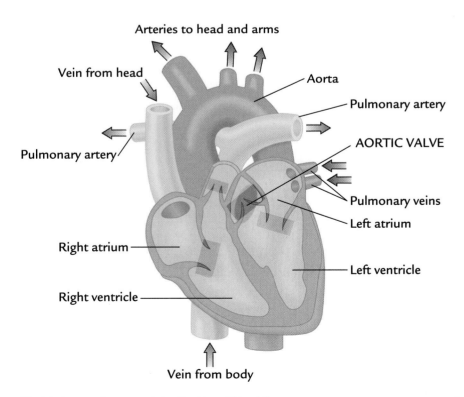

Healthy heart valves control the direction of blood flow.

CHALLENGE

How can you design a heart valve prototype out of common materials?

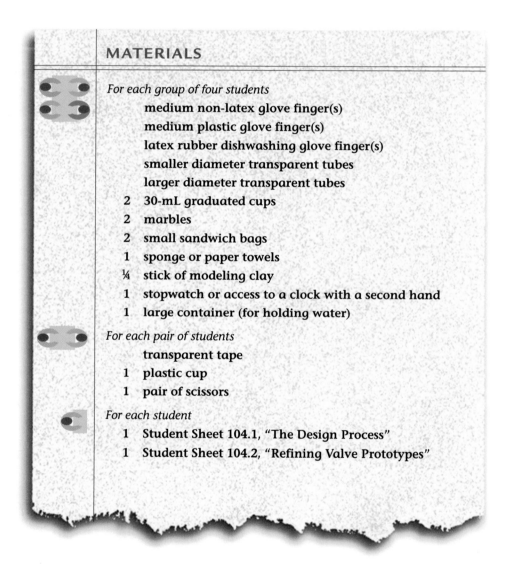

MATERIALS

For each group of four students

    medium non-latex glove finger(s)

    medium plastic glove finger(s)

    latex rubber dishwashing glove finger(s)

    smaller diameter transparent tubes

    larger diameter transparent tubes

2 30-mL graduated cups

2 marbles

2 small sandwich bags

1 sponge or paper towels

¼ stick of modeling clay

1 stopwatch or access to a clock with a second hand

1 large container (for holding water)

For each pair of students

    transparent tape

1 plastic cup

1 pair of scissors

For each student

1 Student Sheet 104.1, "The Design Process"

1 Student Sheet 104.2, "Refining Valve Prototypes"

DESIGNING A PROTOTYPE

As an assignment for your college course in biomedical engineering, you are asked to design a prototype of a heart valve. Your professors want you and your fellow students to learn more about how the heart works while exploring the design process and the advantages of different valve designs. They require that your valve allow fluid to pass quickly in one direction, while allowing less than 30 mL through every 10 seconds in the other direction.

SAFETY

If you are allergic to latex, be sure to tell your teacher before the activity begins. Do not use the latex gloves if you have an allergic reaction to latex.

SAFETY

During this activity, water may spill on the floor, so walk carefully. Be sure to wipe up all spills.

PROCEDURE

Part One: Exploration

1. With your partner, carefully read the Challenge and "Designing a Prototype" (above). Use the materials your teacher gives you to build two different prototype valves that you think will meet these criteria.

2. Test your prototypes to see if they satisfy the design requirements.

3. Discuss with your partner how you could improve the designs of your prototypes.

4. Decide which valve design best meets the design requirements. Draw and label it on Student Sheet 104.1, "The Design Process."

Part Two: Refining Designs

5. Based on the prototype you drew on Student Sheet 104,1, select a factor you could improve in your design. Make this the variable you will test. Then discuss with your partner how you will make a second set of two prototypes to test this variable.

6. Make and test your second set of prototypes.

7. Record your revised designs on Student Sheet 104.2, "Refining Valve Prototypes."

8. Based on your results, make and test a third set of prototypes. Be sure to record your results on Student Sheet 104.2.

Part Three: Sharing and Comparing

9. Prepare to present your best valve prototype to the class during an interactive exhibit.

10. Test other students' valve designs during the interactive exhibition. Take notes on which ones would be most likely to work in the human body. For example, you may want to consider how long a valve would have to last or how a valve would work inside less rigid tubes such as blood vessels.

ANALYSIS

1. **a.** Which of the class's designs best met the design requirements?

   b. What other design requirements would be necessary for a valve to be used in a patient?

   c. Which of the class's designs has the most promise to be developed into an artificial valve for use in patients with the Marfan syndrome? Explain your reasoning.

2. **a.** What factors influenced your design?

   b. What do you think would influence a company designing and marketing an actual heart valve?

3. **a.** How is the design process in this activity similar to other kinds of scientific work?

   b. How is the design process in this activity different from other kinds of scientific work?

4. **Reflection:** What did you learn from this activity about being an inventor?

PROJECT

Engineers design many artificial body parts, including bones. Designing replacement bones for the human body requires understanding the important qualities of bone. Bones are strong enough to support muscles and tissues at rest and during exercise. They weigh surprisingly little, especially in animals that fly. They are also slightly flexible so that, under normal stress, they bend a little rather than break—the way tree branches do. To be useful in a prosthetic arm or leg, artificial bones must also be strong, flexible, and lightweight.

CHALLENGE

How can you design a prototype of an artificial bone that is strong, yet light and flexible?

DESIGN REQUIREMENTS

You are an apprentice to a bioengineer. In preparing to design a bone for a human prosthetic limb, Dr. Chao wants to build a scaled-down model. She provides you with the following design requirements. The model must:

- be no longer or shorter than a drinking straw,

- contain exactly one straw,

- support at least 70 g without crimping,

- have a mass of no more than 7 g, and

- contain no rods, skewers, or wires.

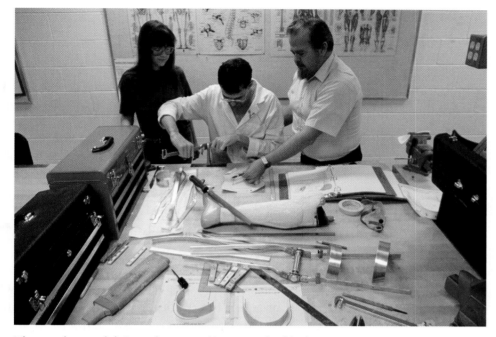

These students and their teacher are making an artificial limb in a community college class.

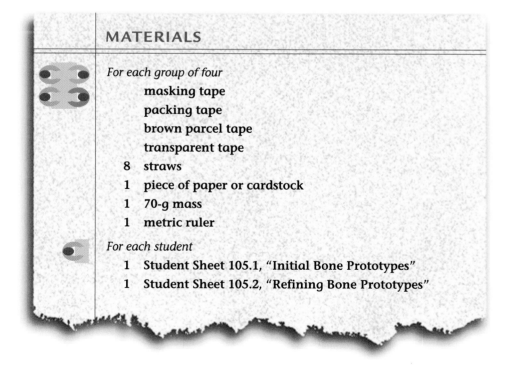

MATERIALS

For each group of four
　　masking tape
　　packing tape
　　brown parcel tape
　　transparent tape
8　straws
1　piece of paper or cardstock
1　70-g mass
1　metric ruler

For each student
1　Student Sheet 105.1, "Initial Bone Prototypes"
1　Student Sheet 105.2, "Refining Bone Prototypes"

PROCEDURE

Part One: Plan

1. Work in your group of four to develop a set of four artificial bone prototypes. Discuss your design ideas and decide which materials you will use. Then decide who will make each prototype.

2. On Student Sheet 105.1, "Initial Bone Prototypes," draw and label diagrams of your bone prototypes. At this point, you do not need to identify the exact amount of each material that you will need in your design.

Part Two: Construct

3. Make the initial prototype assigned to you by your group. Measure or count the amount of each material you are using. **Hint:** Try to use amounts that are easy to measure, such as 10 cm of tape. Measure all materials before you use them.

4. On Student Sheet 105.1, record the amount of each material you used on your diagram.

5. Have all group members share the initial prototypes within your group and make diagrams of your group members' work.

Part Three: Test

6. Measure the mass of each initial prototype. Record the mass on Student Sheet 105.1. Remember, to meet the design criteria, it must be no more than 7 g.

7. Make a mark on each bone 1 cm from one end and another mark 2 cm from the other end.

8. Use the procedure demonstrated by your teacher and shown in Figure 1 below to test your group's initial prototypes' ability to support a 70-g mass. Record the results on Student Sheet 105.1.

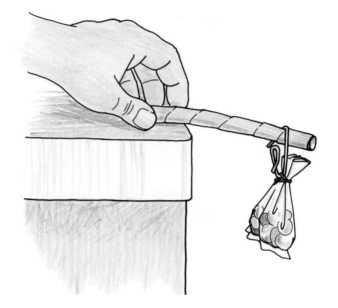

Figure 1: Testing Procedure

Part Four: Redesign and Refine

9. As a group, decide which prototype represents the general type of design you should develop further.

10. Discuss which variable you will investigate and test. For example, you might try varying the amount of tape or the type of tape. This is your experimental variable. You should keep all other variables the same; these are the controlled variables.

11. Exchange ideas about how to test the experimental variable. Develop another set of four prototypes. Draw labeled diagrams of your second set of prototypes on Student Sheet 105.2, "Refining Bone Prototypes." Highlight the experimental variable.

12. Build the four prototypes, working closely with your group members to be sure that the controlled variables are kept the same. Mass the refined prototypes and test them using the procedure shown in Figure 1. Record your results on Student Sheet 105.2.

Part Five: Share Designs

13. As directed by your teacher, present your development process to the class. Include both a description of your procedure for constructing your prototypes and any conclusions you can draw based on the data you collected.

EXTENSION

Test your best prototype to see how much weight is necessary to make it fail. Your prototype fails when it crimps, creases, or bends and does not return to its original straight shape when you remove the weight.

ANALYSIS

1. a. Which of your refined prototypes was strongest?

   b. Which of your refined prototypes was lightest?

   c. Which of the class's artificial bone prototypes looks most promising for future development? Explain.

2. If you had more time and materials, what would your next design look like? Sketch and label it to show what changes you would make.

3. How might a light but strong tube be used, other than to replace bones? List at least three ideas.

4. a. How is the process that you and your group used for this project similar to and different from the process you used in Activity 104, "Designing Artificial Heart Valves"?

   b. How was the design process you just completed similar to and different from designing and conducting scientific experiments?

You built lightweight, strong bone prototypes in Activity 105, "Designing Artificial Bones," but there may be design possibilities that you didn't even consider. One way bioengineers approach problems like this is to observe the structures of body parts that perform similar functions in organisms. By dissecting a chicken wing, you can find out how bird bones can be both lightweight and strong. In this activity, you will also study how the parts of the chicken wing move. This will help you think about how you could design an artificial arm that works in a similar way.

CHALLENGE

How does the structure of an arm or wing affect its function?

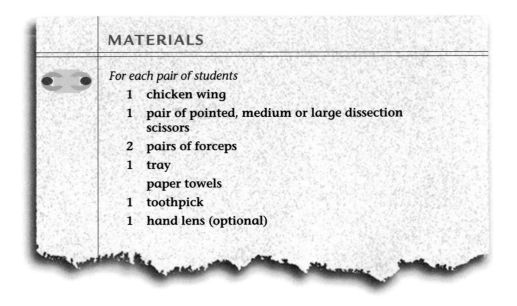

MATERIALS

For each pair of students

1   chicken wing
1   pair of pointed, medium or large dissection scissors
2   pairs of forceps
1   tray
    paper towels
1   toothpick
1   hand lens (optional)

SAFETY

Only one person may dissect at a time—take turns. Keep your fingers out of the way of sharp implements. Do not eat or drink in class. Be very careful not to touch your mouth, nose, or eyes when you are working on the dissection. Wash your hands thoroughly with soap and hot water after completing the dissection.

PROCEDURE

Part One: Comparing the Chicken Wing to the Human Arm

1. Locate the following structures in your arm: shoulder, elbow, and wrist joints; two forearm bones, one upper arm bone, thumb and finger bones.

2. Examine the whole chicken wing.

3. Without cutting yet, feel the wing. Use your fingers to find structures on the chicken wing that are similar to the human arm structures listed in Step 1.

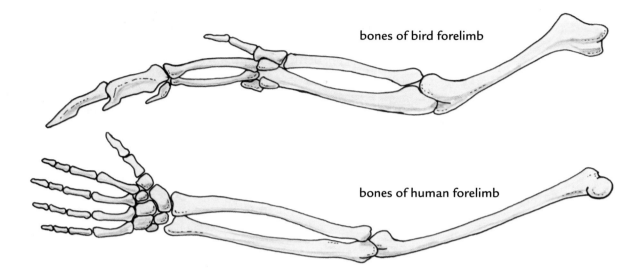

bones of bird forelimb

bones of human forelimb

Part Two: Comparing the Movement of Wings and Arms

4. Turn the wing so the inside is facing up. Use your forceps to pinch up the skin and make a small cut with your scissors, as shown in Figure 1.

Figure 1: Making A Cut
Make a small cut in the skin.

5. As shown in Figure 2, insert the scissors into the cut so that they are parallel to the bones. Be careful that you don't cut through muscle under the skin.

Figure 2: Inserting the Scissors
Insert the tip of the scissors into the small cut.

➤

6. As shown in Figure 3, cut the skin and peel it away from the muscle, using your forceps and scissors to help you. Expose both major joints of the chicken wing.

Observe the tendons, blood vessels, and muscle. Tendons are the shiny strips of tissue that connect muscles to bones.

Figure 3: Cutting the Skin
Cut the skin along the bone, without cutting the muscle.

7. Use your forceps to pull on tendons individually. When muscles contract, they pull on tendons, so when you pull on a tendon, you are modeling the action of a wing muscle (Figure 4). Try to get a part of your chicken wing to "wave" back and forth by pulling on tendons attached to two opposing muscles.

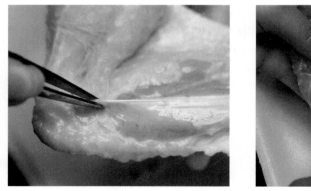

Figure 4: Moving the Chicken Wing
Pulling on the tendon shown in the photo on the left makes the chicken's "hand" move toward the lower "arm," as shown in the photo on the right.

8. Cut through the muscles until one of the chicken's lower arm bones is clearly visible.

9. Break the bone with your fingers. Notice how resistant the bone was to bending. Discuss with your partner whether it is stronger or weaker than the strongest prototype you constructed in the last activity.

10. Examine the inside of the chicken bone. Use a toothpick to explore the texture of the center of the bone, the "marrow."

11. Set the chicken wing out on the tray so that you can see all of the structures.

12. Wash your hands thoroughly with soap and hot water. Don't touch the chicken after you wash.

13. In your science notebook, draw a labeled diagram of the chicken wing. Include the tendons and the structures you located in Step 3.

14. In your notebook, describe what you had to do to make the arm move in opposite directions. Record your observations of the inside of the chicken bone.

15. Follow your teacher's directions for disposing of the chicken wing and for final clean up.

ANALYSIS

1. How are human arms and chicken wings similar? How are they different?

2. What evidence did you find that would help to explain how birds move parts of their wings back and forth? Draw a diagram showing muscles and tendons to help explain your answer.

3. Describe how the structure of bird bones allows them to be both lightweight and strong. Use a diagram to help explain your answer.

4. Now that you know the internal structures of bird bones, would you change your bone prototype from Activity 105? If so, describe how and why. If not, explain why not.

TALKING IT OVER

Scientific discoveries can lead to new inventions, or technologies. As new inventions are used and improved, they often lead to the study and discovery of new science principles. For example, the scientific discovery that glass lenses can be used to magnify objects led to the invention of eyeglasses and telescopes. Telescopes were then used to discover distant planets and stars.

Science and technology are closely related and depend on each other, but they also have distinctive characteristics. Just as there are many kinds of science, there are many fields of study that involve inventions, such as industrial technology, computer technology, and biotechnology. Inventions are designed and built by engineers, scientists, doctors, and a variety of people trying to solve everyday problems.

These high school students are using both tools and ideas to perform a science lab activity.

CHALLENGE

How are science and technology linked and how are they different?

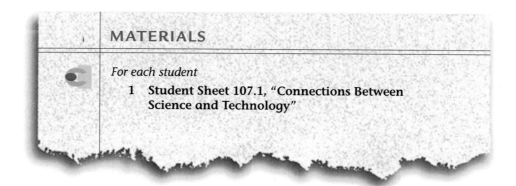

MATERIALS

For each student
1 Student Sheet 107.1, "Connections Between Science and Technology"

PROCEDURE

Part One: Scientists and Inventors

1. Read about the work of three of the people described on pages G33–G36.

2. In your group of four, decide whether you would describe each person as more of a scientist, an inventor, or a combination of the two.

Part Two: Science and Inventions

3. In your group of four, think about the science concepts listed in the center column, Rows 1–3, on Student Sheet 107.1, "Connections Between Science and Technology." In the left-hand column, record any inventions you think have helped to explore these science concepts.

4. Discuss inventions that you think were developed at least partly as a result of the science concepts listed. Record your ideas in the right-hand column.

5. Think of at least two more science concepts that you have studied. Write each in a row of the center column. Try to fill in the left-hand column and the right-hand column as you did in Rows 1–3.

6. **a.** Think of an invention you use in your life. Record this invention in the right-hand column. Think about a science concept that was necessary to understand in order to invent it. Write the science concept in the center column. Try to express the concept in a complete sentence.

   b. Then think about what inventions helped to make the discovery of the science concept possible. Write your ideas in the left-hand column.

ANALYSIS

1. How would you describe the relationship between science and technology? Describe one example from this activity or from your science course that illustrates this relationship.

2. Think about a career that interests you. All careers can benefit from scientific thinking and inventions. Many careers also require some science knowledge. How might your inventing skills and understanding of science help you succeed in your chosen career?

3. Imagine that you could decide how much money a university will provide for scientific research and how much it will provide for development of technology. The university is considering two proposals. One proposal would provide 80% of the funds to scientific research and 20% to technology development. The other proposal would split the funding so that 20% goes to scientific research and 80% goes to technology development.

   Explain whether you would fund one of these two proposals, or whether you would make another proposal. If you would make another proposal, be sure to describe it. Then explain what factors influenced your decision and identify the trade-offs of your choice.

   Hint: To write a complete answer, first state your opinion. Provide two or more pieces of evidence that support your opinion. Then consider all sides of the issue and identify the trade-offs of your decision.

4. **Reflection:** Would you rather be a scientist, an inventor, or a combination of the two? Why did you make this choice? Explain.

EXTENSION 1

Look up other people described in your science book and decide whether you would describe each person as more of a scientist, an inventor, or a combination of the two.

EXTENSION 2

Look up other scientists and inventors working in a field that interests you. You can get started by searching on the internet or by looking in the science pages of newspapers and magazines.

Barbara McClintock (1902–1992)

McClintock studied how genes are transmitted from parent to offspring and made an unexpected discovery. She demonstrated that genes can move from one chromosome to another and that this can change the function of the gene. For many years, the scientific community did not realize the significance of her work. She received the Nobel Prize in Medicine in 1983.

Lloyd Hall (1884–1971)

Hall was an African-American researcher who earned a Ph.D. in chemistry and was awarded over 100 patents. Many of the patents were issued for new preservatives that helped to prevent food from spoiling. His methods are still widely used today. After his retirement, he became a consultant to the Food and Agriculture Organization of the United Nations.

David Ho (1952-   )

Ho is the director of the Aaron Diamond AIDS Research Center in New York City. He is one of the pioneers who discovered how the virus is (and is not) transmitted. Ho is known for developing an approach that provides anti-AIDS drugs to people as soon as possible after they contract the virus, rather than waiting for symptoms to appear. In 1996 he was named *Time* magazine's Person of the Year.

Helen Murray Free (1923–)

During her long career as a research chemist with Miles Laboratories and Bayer, Inc., Free has obtained several patents for her improvements in medical testing. For example, she developed the first home "dip and read" urine test that allowed diabetics to check their own blood sugar levels. In 1993, she served as President of the American Chemical Society.

Lydia Villa–Komaroff (1947–)

Villa-Komaroff was a member of the research team at MIT that demonstrated that bacterial cells can be used to produce insulin to treat people with diabetes. This was accomplished by inserting the human insulin gene into the DNA of the bacteria. This discovery helped launch recombinant DNA technology. Villa-Komaroff is a founding member of the Society for the Advancement of Chicanos and Native Americans in Science.

Edward O. Wilson (1929–)

Wilson grew up in Alabama and went to college there. He is currently Emeritus Professor of Entomology—the study of insects—at Harvard. He is known for his controversial comparisons between ant behavior and human behavior. Currently he studies and writes about the diversity of life, suggesting that it is worth protecting not only for its usefulness to humans, but also for its own sake.

Charles Drew (1904–1950)

Drew was a doctor and researcher who studied the properties of blood, especially plasma. He developed methods for storing plasma for long periods of time and invented blood banks and bloodmobiles. He directed the first Red Cross Blood Bank and become head of the department of surgery at Howard University and chief surgeon at Freedman's Hospital.

Bessie Blount (1914–?)

Blount was an African-American physical therapist who worked with patients who had been disabled during World War II. She invented devices that allowed patients who had lost their arms to eat independently. She received a patent for one of her devices in 1951, but was unable to convince the Veterans' Administration to use the device in its hospitals. Few details are known about her life, despite her initiative and useful work.

Jane Goodall (1934–   )

Goodall has spent most of her life collecting information about how chimpanzees live and interact. She conducted much of her research in the field, living for long periods of time on Gombe Stream in Tanzania. She founded the Jane Goodall Institute for Wildlife Research, Education, and Conservation, and she has published many articles and several books about chimp behavior.

Wilson Greatbatch (1919–   )

Greatbatch invented the first cardiac pacemaker that could be implanted in patients. He was trained as an electrical engineer. He has patented many inventions, including a compact and long-lasting battery for pacemakers. In 1986, he was inducted into the National Inventors Hall of Fame.

Stephen Jay Gould (1941-   )

Gould is a paleontologist and evolutionary biologist at Harvard University. He was one of the first to suggest that species evolve slowly until fairly rapid changes lead to the formation of new species. Though his own work focuses on the evolution of land snails, he is the author of many books on a broad range of topics related to evolution. He wrote a monthly column in *Natural History* magazine for 30 years.

Lynn Margulis (1938–   )

Margulis is a cell and microbial biologist who has researched the evolution of cells and cell organelles such as chloroplasts and mitochondria. She is well-known for the idea that these organelles in plant and animal cells evolved from simple cells similar to bacteria. She is a professor at the University of Massachusetts and has published numerous research papers and several books.

PROJECT

Humans not only design technological tools; they also provide the model or inspiration for many tools. For instance, robotic arms are useful not only to replace human arms, but also to work with hazardous materials or in remote environments—such as handling radioactive materials or gathering samples from the ocean depths or from other planets. Brain and heart surgeons use small robotic arms for very delicate operations, and these arms generally require computer technology. Surgeons also use robotic arms in new "keyhole" techniques, in which small incisions are used for major operations, including heart surgery. You can probably think of many other everyday uses.

This model shows the mechanical arm of the Mars Viking lander collecting soil samples.

CHALLENGE

How can you make a mechanical arm that moves a mass back and forth?

➢

DEVELOPING A MECHANICAL ARM

You are a mechanical engineer, and today you face an unusual task. Your supervisor, Dr. Garcia, has been given the task of developing a mechanical arm for handling dangerous substances. He has to attend a conference and assigns the development of a prototype to you. He tells you that the arm must be able to move an object with a mass of about 70 grams 6 cm in two opposite directions. You can use only one of your hands to operate the controls. You must use your other hand to hold the mechanical arm in place.

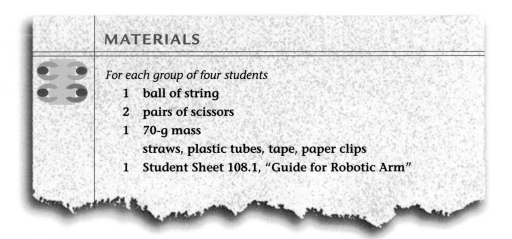

MATERIALS

For each group of four students
- 1   **ball of string**
- 2   **pairs of scissors**
- 1   **70-g mass**
-     **straws, plastic tubes, tape, paper clips**
- 1   **Student Sheet 108.1, "Guide for Robotic Arm"**

PROCEDURE

1. As a group, carefully read "Developing a Mechanical Arm" (above). Determine what the design requirements are, and make a list of them in your science notebook.

2. In your group, discuss and record your ideas about how you will construct your prototypes.

3. In your science notebook, create a data table on which you can record your observations and collect your data.

4. As a group, build and test your first prototype(s). (Your group may wish to make more than one at a time.) Remember that you may move only one of your hands to control the mechanical arm. The other hand and arm must be still and hold the mechanical arm in a fixed position. Record your observations and test results in your data table.

5. Discuss your results with your group. Decide which features of your prototype(s) were helpful and which were not. Record ideas for improving your best prototype. If possible, focus on testing the effect of one variable.

6. Make and test one or more improved prototypes. Record your results in your data table.

7. Present your best prototype to the class: demonstrate its function and describe your design process.

ANALYSIS

1. In what real-world applications could robotic arms be useful? Include everyday as well as high-tech examples.

2. What are some of the trade-offs of inventing as a group compared to working individually?

3. Thomas Edison invented over 1,000 inventions, including the movie camera, the telegraph, and the phonograph (record player). He once said, "It is what you do after failure that counts." What did he mean? Explain your answer with an example from your work during this unit.

4. Isaac Newton is thought to have said, "If I have seen a little farther than others, it is because I have stood on the shoulders of giants." What did he mean? Explain your answer with a real example of an invention or scientific discovery.

5. **Reflection:** Describe the relationship between science and technology.

Index

A **bold** page number identifies the page on which the term is defined.

Credits

Abbreviations: t (top), m (middle), b (bottom), l (left), r (right), c (center)

All illustrations by Seventeenth Street Studios, except as follows: Pages B26, B36, B43, B63, B66, B72, B73, B80, B81, G-15: Precision Graphics.

"Talking It Over" icon photo: ©Michael Keller/The Stock Market

Cover photo (kids running): ©2001 David Young-Wolff/Stone

Unit A

Unit opener (A-2, A-3): bl: ©S. Fraser/Photo Researchers, Inc.; bc: ©2001 Richard Price/FPG; br: Donna Markey; tl: ©2001 Charles Thatcher/Stone; tc: ©2001 B. Busco/The Image Bank.

A-8 ©2001/The Image Bank; A-9 ©Bettmann/CORBIS; A-25 ©Digital Art/CORBIS; A-34 Heart Pacemaker ©Department of Clinical Radiology, Salisbury District Hospital/SPL/Photo Researchers, Inc.; A-44 photo courtesy of the World Health Organization; A-46 ©Laura Dwight/DoctorStock.com.

Unit B

Unit opener (B-2): bl: Donna Markey.

B-37 (photo) ©Martin Dohrn/Royal College of Surgeons/SPL/Photo Researchers, Inc.; B-69 © Helen King/Corbis; B-70 courtesy of the Provident Foundation, Chicago, Illinois; B-71 © Shelley D. Spray/Corbis; B-72 © Ted Spiegel/Corbis; B-80 (photo) ©SPL/Photo Researchers, Inc.

Unit C

Unit opener (C-2, C-3): tm: ©Charles O'Rear/CORBIS; lb: ©Science Pictures Limited/CORBIS; mb: ©Premium Stock/CORBIS; m: ©Ed Eckstein/CORBIS; rb: ©Jack Fields/CORBIS; rm: Dennis Kunkel/PHOTOTAKE

C-4 ©2001 Stuart McClymont/Stone; C-8 Image courtesy from the Centers for Disease Control with permission from Christy Turlington; C-10 Michael Brill, Louisville, Kentucky; C-12 Will & Deni McIntyre/Photo Researchers, Inc.; C-18 ©Richard T. Nowitz/PHOTOTAKE; C-19 Image courtesy of TDR image library, TDR Communications and the World Health Organization; C-20 Philip Gould/CORBIS; C-22 l: ©2001 Stewart Cohen/Stone, r: Dr. Dennis Kunkel/PHOTOTAKE; C-27 tl: ©M.I. Walker/Photo Researchers, Inc., tm, tr, bl, bm: ©Barry Runk/Stan/Grant Heilman Photography, br: ©Eric Grave/PHOTOTAKE; C-28 ©Barry Runk/Stan/Grant Heilman Photography; C-32 ml: ©Bettmann/CORBIS, bl: ©Charles O'Rear/CORBIS, br: ©Lester V. Bergman/CORBIS; C-33 tl: ©Bettmann/CORBIS, mr: ©Science Pictures Limited/CORBIS; C-34 tl: Sue Boudreau, ml: ©Science Pictures Limited/CORBIS, bl: Dr. Dennis Kunkel/PHOTOTAKE; C-35 "Leo the cat": Sylvia Parisotto; C-36 Dr. Dennis Kunkel/PHOTOTAKE;

C-38 ml: ©Bettman/CORBIS, bl: ©Bettmann/CORBIS; C-39 ©G. Watson/Photo Reseachers, Inc.; C-42 ©Bettmann/CORBIS; C-47 ©2001 Tracy Frankel/The Image Bank; C-50 ©Barry Runk/Stan/Grant Heilman Photography; C-51 ©Jim Zuckerman/CORBIS; C-58 LesterV. Bergman/CORBIS; C-60 ©Dr. Dennis Kunkel/PHOTOTAKE; C-62 tl: ©CNRI/PHOTOTAKE, tr: ©CNRI/Photo Researchers, Inc., bl: ©Lester V. Bergman/ CORBIS, br: ©Tina Carvalho; C-64 tl, tr, bl: ©Eric Grave/PHOTOTAKE, mr: ©Carolina Biological Supply Co./PHOTOTAKE; C-77 r: ©Carolina Biological Supply Co./ PHOTOTAKE, l: ©NCI/Photo Researchers, Inc.; C-80 ©Dr. Dennis Kunkel/PHOTOTAKE; C-82 ©Bruce Iverson/DoctorStock.com; C-89 "Guidelines for Doctors Prior to Surgery" source: Thomas Barber, MD, "Guidelines for Food Industry Workers" source: Utah Department of Health; C-90 ©2001 Richard Price/FPG; C-93 ©CORBIS; C-97 ©C. James Webb/PHOTOTAKE; C-107 ©Will & Deni McIntyre/Photo Researchers, Inc.

Unit D

Unit opener (D-2, D-3): tm: ©2001 Michael Krasowitz/FPG; tl: ©2001 Bob Elsdale/ The Image Bank; bl: ©2001 Doug Struthers/Stone; mr: ©2001 Charles Thatcher/Stone; bc: ©Lester V. Bergman/CORBIS; br: Donna Markey

D-4 ©2001 David Young-Wolff/Stone; D-9 ©Tania Midgley/CORBIS; D-16 © Buddy Mays/CORBIS; D-18 ©Lester V. Bergman/CORBIS; D-19 t: ©2001 Spike Walker/Stone, b: Sylvia Parisotto; D-20 ©Dr. Dennis Kunkel/PHOTOTAKE; D-24 ©2001 American Images Inc./FPG; D-34 ©Bettmann/CORBIS; D-44 © Carolina Biological Supply Co./ PHOTOTAKE; D-47 ©2001 Spike Walker/Stone; D-49 ©2001 Spike Walker/Stone; D-51 ©Biophoto Associates/Photo Researchers, Inc.; D-55 m: ©Lynda Richardson/ CORBIS, bl: © Bob Krist/CORBIS, br: ©2001 Steve Satushek/The Image Bank; D-64 Pedigree adapted with permission from Robert J. Huskey, Emeritus Professor, Dept. of Biology, University of Virginia; D-78 ©David & Peter Turnley/CORBIS; D-87 ©Richard T. Nowitz/CORBIS; D-91 ©Bettmann/CORBIS

Unit E

Unit opener (E-2, E-3): tl: ©Lynda Richardson/CORBIS; tm: Sylvia Parisotto; bl: ©Dan Guravich/CORBIS; m: ©Australian Picture Library/CORBIS; mr: Sylvia Parisotto; bm: ©Anna Clopet/CORBIS; br: Dr. Herbert Thier

E-5 ©Liam Dale by permission LDTV, England; E-6 ©Frank Lane Picture Agency/CORBIS; E-10 ©S. van Mechelen, courtesy of the Exotic Species Graphics Library; E-12 t: ©Buddy Mays/CORBIS, b: courtesy of Jack Leonard, New Orleans Mosquito Control Board; E-13 t: © O. Alamany & E. Vicens/CORBIS, b: California Department of Food and Agriculture; E-14 b: Martha L. Walter, Michigan Sea Grant; E-15 ©Lynda Richardson/ CORBIS; E-16 ©Joel W. Rogers/CORBIS; E-23 tl: Dr. Herbert Thier, tm: ©W. Wayne Lockwood, M.D./CORBIS, tr: ©Stephen Frink/CORBIS, bl: ©Michael & Patricia Fogden/ CORBIS; E-25 t: ©Papilio/CORBIS, b: ©Buddy Mays/CORBIS; E-28: l: ©Brandon D. Cole; E-29 ©Neil Rabinowitz/CORBIS; E-32 ©Morton Beebe, S.F./CORBIS; E-33 ©Neil Rabinowitz/ CORBIS; E-43 GLSGN Exotic Species Library; E-44 ©Frank Lane Picture Agency/CORBIS; E-47 map courtesy of the U.S. Geological Survey; E-52 ©AFP/CORBIS; E-59 ©Gary Braasch/ CORBIS; E-67 GLSGN Exotic Species Library; E-72 ©2001 Ben Osborne/Stone; E-80 Stephen Stewart, Michigan Sea Grant; E-81 Ron Peplowski, Detroit Edison, Monroe Michigan Power Station; E-83 Steve Krynock; E-85 ©Kevin Fleming/CORBIS

Unit F

Unit opener (F-2, F-3): tl: Roberta Smith; tm: ©Charles Mauzy/CORBIS; bl: Roberta Smith; m: ©Kevin Schafer/CORBIS; mr: ©Jonathan Blair/CORBIS; bm: ©Lester V. Bergman/ CORBIS; br: Donna Markey

F-4 tl: ©C.Iverson/Photo Researchers, Inc., tr: ©Bettmann/CORBIS, br: ©Kevin Fleming/ CORBIS; F-7 tr: ©2001Manoj Shah/Stone F-14 ©Annie Griffiths Belt/CORBIS; F-15 ©Francesc Muntada/CORBIS; F-19 ©Kevin Schafer/CORBIS; F-21 ©2001 Lori Adamski Peek/Stone; F-22 ©Charles Mauzy/CORBIS; F-37 br: ©W. Perry Conway/CORBIS; F-41 bl: ©Gary W. Carter/CORBIS, br:©George Lepp/CORBIS; F-46 ©F. McConnaughey/ Photo Researchers, Inc.; F-49 Images courtesy of Dr. Robert Rothman; F-50 l: ©Papilio/ CORBIS, r: ©Sea World, Inc./Corbis; F-67 br: ©Hulton-Deutsch Collections/CORBIS; F-69 l: ©2001 Cesar Lucas Abreu/The Image Bank, r: Annie Griffiths Bell/CORBIS; F-70 ©David & Peter Turnley/CORBIS

Unit G

Unit opener (G-2, G-3): tl: ©Layne Kennedy/CORBIS; m: ©2001 Flip Chalfant/ The Image Bank

G-4 ©Kevin R. Morris/CORBIS; G-7 John Quick, photographed for SEPUP; G-11 Aimee Mullins as quoted in Shepard, E. "Confidence is the Sexiest Thing a Woman Can Have." Parade, June 21,1998, image: ©Roger Ressmeyer/CORBIS; G-12 ©1986 Jeff Smoot; G-21 ©Bob Rowan; Progressive Image/CORBIS G-27, G-28 images by John Quick, photographed for SEPUP; G-30 ©Robert Maass/CORBIS; G-33 t: ©Bettmann/CORBIS, b: ©AFP/CORBIS; G-34 t: courtesy of Northwestern University, b: ©AP Photo; G-35 ©Kennan Ward/CORBIS; G-36 ©Wally McNamee/CORBIS; G-37 ©Lowell Georgia/ CORBIS